The Four-Category Ontology

A Metaphysical Foundation for Natural Science

E. J. LOWE

CLARENDON PRESS · OXFORD

This book has been printed digitally and produced in a standard specification in order to ensure its continuing availability

OXFORD
UNIVERSITY PRESS

Great Clarendon Street, Oxford OX2 6DP
Oxford University Press is a department of the University of Oxford.
It furthers the University's objective of excellence in research, scholarship,
and education by publishing worldwide in
Oxford New York

Auckland Cape Town Dar es Salaam Hong Kong Karachi
Kuala Lumpur Madrid Melbourne Mexico City Nairobi
New Delhi Shanghai Taipei Toronto
With offices in
Argentina Austria Brazil Chile Czech Republic France Greece
Guatemala Hungary Italy Japan South Korea Poland Portugal
Singapore Switzerland Thailand Turkey Ukraine Vietnam

Oxford is a registered trade mark of Oxford University Press
in the UK and in certain other countries

Published in the United States
by Oxford University Press Inc., New York

© E. J. Lowe 2006

The moral rights of the author have been asserted

Database right Oxford University Press (maker)

Reprinted 2009

All rights reserved. No part of this publication may be reproduced,
stored in a retrieval system, or transmitted, in any form or by any means,
without the prior permission in writing of Oxford University Press,
or as expressly permitted by law, or under terms agreed with the appropriate
reprographics rights organization. Enquiries concerning reproduction
outside the scope of the above should be sent to the Rights Department,
Oxford University Press, at the address above

You must not circulate this book in any other binding or cover
And you must impose this same condition on any acquirer

ISBN 978-0-19-922981-9

THE FOUR-CATEGORY ONTOLOGY

A Metaphysical Foundation for Natural Science

E. J. Lowe sets out and defends his theory of what there is. His four-category ontology is a metaphysical system that recognizes two fundamental categorial distinctions which cut across each other to generate four fundamental ontological categories. The distinctions are between the particular and the universal and between the substantial and the non-substantial. The four categories thus generated are substantial particulars, non-substantial particulars, substantial universals and non-substantial universals. Non-substantial universals include properties and relations, conceived as universals. Non-substantial particulars include property-instances and relation-instances, otherwise known as non-relational and relational tropes or modes. Substantial particulars include propertied individuals, the paradigm examples of which are persisting, concrete objects. Substantial universals are otherwise known as substantial kinds and include as paradigm examples natural kinds of persisting objects.

This ontology has a lengthy pedigree, many commentators attributing it to Aristotle on the basis of certain passages in his apparently early work, the *Categories*. At various times during the history of Western philosophy, it has been revived or rediscovered, but it has never found universal favour, perhaps on account of its apparent lack of parsimony as well as its commitment to universals. In pursuit of ontological economy, metaphysicians have generally preferred to recognize fewer than four fundamental ontological categories. However, Occam's razor stipulates only that we should not multiply entities beyond necessity; Lowe argues that the four-category ontology has an explanatory power unrivalled by more parsimonious systems, and that this counts decisively in its favour. He shows that it provides a powerful explanatory framework for a unified account of causation, dispositions, natural laws, natural necessity and many other related matters, such as the semantics of counterfactual conditionals and the character of the truthmaking relation. As such, it constitutes a thoroughgoing metaphysical foundation for natural science.

E. J. Lowe is Professor of Philosophy at the University of Durham.

Preface

By 'the four-category ontology' I mean a system of ontology which recognizes two fundamental categorial distinctions which cut across each other to generate four fundamental ontological categories, these distinctions being between the *particular* and the *universal* and between the *substantial* and the *non-substantial*. The four categories thus generated are substantial particulars, non-substantial particulars, substantial universals, and non-substantial universals. Non-substantial universals include properties and relations, conceived as universals. Non-substantial particulars include property- and relation-instances, otherwise known as non-relational and relational 'tropes' or—as I prefer to call them—*modes*. Substantial particulars include propertied individuals, the paradigm examples of which are persisting, concrete objects. Substantial universals are otherwise known as *substantial kinds* and include as paradigm examples natural kinds of persisting objects. This ontology has a lengthy pedigree, many commentators attributing it to Aristotle on the basis of certain passages in his apparently early work, the *Categories*. At various times during the history of western philosophy, it has been revived or rediscovered, but it has never found widespread favour, perhaps on account of its apparent lack of parsimony as well as its commitment to universals. In pursuit of ontological economy, metaphysicians have generally preferred to recognize fewer than four fundamental ontological categories. Indeed, present-day metaphysicians who are self-styled 'trope theorists' advertise their position as being a one-category ontology and see this as being one of its principal attractions. But, of course, Occam's razor stipulates only that we should not multiply entities (or, more generally, ontological distinctions) *beyond necessity* and it may be argued—as I do indeed argue in the course of this book—that the four-category ontology has an explanatory power which is unrivalled by more parsimonious systems and that this counts decisively in its favour.

In earlier work, I have presented and defended various aspects of the four-category ontology and shown how it provides a powerful explanatory framework for a unified account of causation, dispositions, natural laws, natural necessity, and other related matters, such as the semantics of counterfactual conditionals. Although I first explicitly announced my allegiance to the four-category ontology in Chapter 9 of my recent book, *The Possibility of Metaphysics: Substance, Identity, and Time* (Oxford: Clarendon Press, 1998), much of the framework and some of its applications were already in place in my earlier book, *Kinds of Being: A Study of Individuation, Identity, and the Logic of Sortal Terms* (Oxford: Blackwell, 1989). What was crucially lacking in that earlier book was an adequate recognition of the category of non-substantial particulars, that is, of property- and relation-instances, or 'modes'. Since the publication of *The Possibility of Metaphysics*,

I have written a number of papers repairing this omission and developing the theory. It is this new work that I am now bringing together and extending to produce, in the form of this book, a systematic and comprehensive account of the four-category ontology and its many explanatory applications. Some chapters of the book are accordingly based upon, or draw upon, papers of mine that have been published during the last few years, although much else is entirely new. In particular, Chapters 1, 2, 3, 4, 6, 8, 9, and 11 draw upon the following papers respectively, and I am grateful to the publishers and editors concerned for permission to use the material in question in this way: 'Recent Advances in Metaphysics', *Facta Philosophica* 5 (2003), pp. 3–24; 'A Defence of the Four-Category Ontology', in C. U. Moulines and K. G. Niebergall (eds), *Argument und Analyse* (Paderborn: Mentis, 2002), pp. 225–40; 'Some Formal Ontological Relations', *Dialectica* 58 (2004), pp. 297–316; 'Syntax and Ontology: Reflections on Three Logical Systems', in D. S. Oderberg (ed.), *The Old New Logic: Essays on the Philosophy of Fred Sommers* (Cambridge, MA: MIT Press, 2005); 'Properties, Modes, and Universals', *The Modern Schoolman* 74 (2002), pp. 137–50; 'Dispositions and Laws', *Metaphysica* 2 (2001), pp. 5–23; 'Kinds, Essence, and Natural Necessity', in A. Bottani, M. Carrara and P. Giaretta (eds), *Individuals, Essence, and Identity: Themes of Analytic Metaphysics* (Dordrecht: Kluwer, 2002), pp. 189–206; and 'Metaphysical Realism and the Unity of Truth', in A. Bächli and K. Petrus (eds), *Monism* (Frankfurt: Ontos Verlag, 2003), pp. 109–23. In addition, sections 7.8 and 8.12 are based, respectively, on my 'The Particular–Universal Distinction: A Reply to MacBride', *Dialectica* 58 (2004), pp. 335–40 and my review of George Molnar's *Powers: A Study in Metaphysics*, *British Journal for the Philosophy of Science* 55 (2004), pp. 817–22. Finally, an earlier version of Chapter 12 was written for a conference on Truths and Truthmakers, held in Aix-en-Provence in 2004, and in that form will appear in the published proceedings of the conference, *Metaphysics and Truthmakers*, ed. J.-M. Monnoyer (Frankfurt: Ontos Velag). I am also extremely grateful to Peter Momtchiloff of Oxford University Press for all his encouragement and assistance, as well as to two anonymous readers for the Press for their very helpful comments on the penultimate version of the manuscript. My debts to other people are too numerous to list, but in many cases are exhibited in my references to their published work. However, I should like to record my particular gratitude to John Heil for many fruitful discussions on the subject of fundamental ontology during the last ten years or so. Finally, I want to express my warm thanks to the British Academy and the Leverhulme Trust for awarding me a Senior Research Fellowship for the academic year 2003–4, which enabled me to complete work on the manuscript.

Let me close this preface with a few remarks about the structure of the book. It is divided into four parts—the first setting out the framework of the four-category ontology, the second focusing on its central distinction between object and property, the third exploring its applications in the philosophy of natural science, and the fourth dealing with fundamental issues of truth and realism. Because the book's concerns are interrelated in numerous and complex ways, I have considered it appropriate to remind the reader in various places of key features of the four-category ontology, deeming a certain amount of repetition preferable to reliance on frequent cross-reference from one part of the book to another. An additional advantage of this method of proceeding is that readers whose main interest lies in just one part of the book are not compelled to read others as well.

Contents

Preface — v
List of Figures — xiii
List of Tables — xiv

PART I: METAPHYSICS, ONTOLOGY, AND LOGIC

Chapter 1: Ontological Categories and Categorial Schemes — 3
1.1 Philosophy, metaphysics, and ontology — 3
1.2 Ontological categories — 5
1.3 Some competing ontological systems — 8
1.4 States of affairs and the truthmaker principle — 11
1.5 Laws of nature and properties as ways of being — 13
1.6 The four-category ontology — 15

Chapter 2: The Four-Category Ontology and its Rivals — 20
2.1 Central principles of the four-category ontology — 20
2.2 Properties as universals and properties as particulars — 23
2.3 Substantial universals and substantial particulars — 25
2.4 In defence of universals — 28
2.5 Dispositional versus occurrent predication — 30
2.6 Advantages of the four-category ontology — 32

Chapter 3: Some Formal Ontological Relations — 34
3.1 Dependence relations and their foundation — 34
3.2 Ontological categories and their organization — 38
3.3 The ontological status of the categories — 40
3.4 Are formal ontological relations elements of being? — 44
3.5 Form and content in ontology — 47
3.6 Composition and constitution — 49

Chapter 4: Formal Ontology and Logical Syntax — 52
4.1 Traditional formal logic versus quantified predicate logic — 52
4.2 The two-category ontology of Frege–Russell logic — 56
4.3 How two distinctions generate four categories — 57
4.4 The Ontological Square — 60
4.5 Sortal logic and the four-category ontology — 62

PART II: OBJECTS AND PROPERTIES

Chapter 5: The Concept of an Object in Formal Ontology — 69
- 5.1 Objects, things, and entities — 69
- 5.2 Objects and properties — 70
- 5.3 Ramsey's problem — 72
- 5.4 The individuality of objects — 75
- 5.5 Particulars and universals — 76
- 5.6 Aristotle's four-category ontology reconstituted — 78
- 5.7 Events and processes — 80
- 5.8 Numbers and other abstract objects — 81
- 5.9 Frege on objects and concepts — 83
- 5.10 The ontological status of concepts — 85

Chapter 6: Properties, Modes, and Universals — 87
- 6.1 Properties and predicates — 87
- 6.2 Universals and particulars — 88
- 6.3 Ways of being — 90
- 6.4 Instantiation versus characterization — 91
- 6.5 The four-category ontology revisited — 93
- 6.6 Exemplification and the copula — 95
- 6.7 Modes versus tropes and transcendent universals — 96
- 6.8 Two conceptions of immanence — 98

Chapter 7: Ramsey's Problem and its Solution — 101
- 7.1 The challenge of Ramsey's problem — 101
- 7.2 Ramsey's conflation of two different distinctions — 102
- 7.3 Ramsey's objections to the universal–particular distinction — 104
- 7.4 The lesson of Ramsey's arguments — 107
- 7.5 The advantages of a substance ontology — 109
- 7.6 The status and basis of the universal–particular distinction — 110
- 7.7 The dissolution of Ramsey's problem — 112
- 7.8 A response to MacBride's Ramseian objections to the four-category ontology — 113

PART III: METAPHYSICS AND NATURAL SCIENCE

Chapter 8: Dispositions and Natural Laws — 121
- 8.1 Dispositional versus categorical properties — 121
- 8.2 Predicates and properties — 122

8.3	The four-category ontology: a brief résumé	123
8.4	Dispositional versus occurrent predication	124
8.5	The ontological ground of the dispositional–occurrent distinction	125
8.6	Laws of nature	127
8.7	Natural powers and conditional statements	129
8.8	Laws and universals	130
8.9	Nomic necessity	132
8.10	A comparison with Martin's account	133
8.11	Why universals?	134
8.12	Some comments on Molnar's theory of powers	136

Chapter 9: Kinds, Essence, and Natural Necessity — 141

9.1	Natural necessity and metaphysical necessity	141
9.2	The logical form of law statements	142
9.3	Replies to objections	145
9.4	Laws, counterfactuals, and natural necessity	146
9.5	Scientific essentialism and the identity conditions of properties	149
9.6	How is knowledge of laws possible?	152

Chapter 10: Categorial Ontology and Scientific Essentialism — 156

10.1	Laws and necessity revisited	156
10.2	Natural laws and natural kinds	158
10.3	Causal powers and liabilities	159
10.4	Comparisons with other views	161
10.5	Laws and scientific essentialism	163
10.6	Does water *necessarily* dissolve sodium chloride?	165
10.7	Bradley's regress and the ontological structure of laws	166
10.8	How some laws may be contingent	169
10.9	Structure-based laws	171
10.10	The questionable notion of nomic necessity	172

PART IV: TRUTH, TRUTHMAKING, AND METAPHYSICAL REALISM

Chapter 11: Metaphysical Realism and the Unity of Truth — 177

11.1	Metaphysical realism and alethic monism	177
11.2	Truth-bearers and their multiplicity	177

11.3	The ineliminability of truth	180
11.4	Truthmakers and their multiplicity	182
11.5	Facts and truthmaking	184
11.6	The principle of non-contradiction and the indivisibility of truth	188
11.7	The ontological implications of relativism	190

Chapter 12: Truthmaking, Necessity, and Essential Dependence — 192

12.1	Truth and truthmaking	192
12.2	Formal ontological predicates	193
12.3	Ontology, categories, and metaphysical realism	195
12.4	Identity, essence, and essential dependence	198
12.5	The varieties of metaphysical dependence	200
12.6	Truthmaking as essential dependence	202
12.7	Why facts are not needed as truthmakers	204
12.8	How are contingent truths possible?	205
12.9	Propositions and what they are 'about'	207
12.10	A sketch of a theory of truthmaking	209

Bibliography — 211
Index — 217

List of Figures

Fig. 1.1	A fragment of the hierarchy of categories	8
Fig. 1.2	The four-category ontology	18
Fig. 2.1	The Ontological Square (I)	22
Fig. 3.1	Topmost levels of the categorial hierarchy	39
Fig. 3.2	The Ontological Square (II)	40
Fig. 4.1	The Ontological Square (III)	60
Fig. 5.1	The Ontological Square (IV)	79
Fig. 6.1	The Ontological Square (V)	93
Fig. 7.1	The four-category ontology	111
Fig. 7.2	Patterns of ontological dependency between the four basic categories	117
Fig. 8.1	The ontological ground of the dispositional–occurrent distinction	126

List of Tables

Table 1.1	Four ontological systems	11
Table 2.1	Advantages of the four-category ontology	32

PART I

METAPHYSICS, ONTOLOGY, AND LOGIC

1

Ontological Categories and Categorial Schemes

1.1 PHILOSOPHY, METAPHYSICS, AND ONTOLOGY

There is a widespread assumption amongst non-philosophers, which is shared by a good many practising philosophers too, that 'progress' is never really made in philosophy, and above all in metaphysics. In this respect, philosophy is often compared, for the most part unfavourably, with the empirical sciences, and especially the natural sciences, such as physics, chemistry and biology. Sometimes, philosophy is defended on the grounds that to deplore the lack of 'progress' in it is to misconceive its central aim, which is to challenge and criticize received ideas and assumptions rather than to advance positive theses. But this defence itself is liable to be attacked by the practitioners of other disciplines as unwarranted special pleading on the part of philosophers, whose comparative lack of expertise in other disciplines, it will be said, ill-equips them to play the role of all-purpose intellectual critic. It is sometimes even urged that philosophy is now 'dead', the relic of a pre-scientific age whose useful functions, such as they were, have been taken over at last by genuine sciences. What were once 'philosophical' questions have now been transmuted, allegedly, into questions for more specialized modes of scientific inquiry, with their own distinctive methodological principles and theoretical foundations.

This dismissive view of philosophy is at once shallow and pernicious. It is true that philosophy is not, properly speaking, an empirical science, but there are other disciplines of a non-empirical character in which progress most certainly can be and has been made, such as mathematics and logic. So there is no reason, in principle, why progress should not be made in philosophy. However, it must be acknowledged that even professional philosophers are in much less agreement amongst themselves as to the nature of their discipline and the proper methods of practising it than are mathematicians and logicians. There is more disagreement about fundamentals in philosophy than in any other area of human thought. But this should not surprise us, since philosophy is precisely concerned with the most fundamental questions that can arise for the human intellect.

The conception of philosophy that I favour is one which places metaphysics at the heart of philosophy and ontology—the science of being—at the heart of metaphysics.[1] Why do we need a 'science of being', and how is such a science possible? Why cannot each special science, be it empirical or *a priori*, address its own ontological questions on its own behalf, without recourse to any overarching 'science of being'? The short answer to this question is that reality is one and truth indivisible. Each special science aims at truth, seeking to portray accurately some part of reality. But the various portrayals of different parts of reality must, if they are all to be true, fit together to make a portrait which can be true of reality as a whole. No special science can arrogate to itself the task of rendering mutually consistent the various partial portrayals: that task can alone belong to an overarching science of being, that is, to ontology. But we should not be misled by this talk of 'portraits' of reality. The proper concern of ontology is not the portraits we construct of it, but reality itself.

Here, however, we encounter one of the great divides in philosophy, whose historical roots lie in the seventeenth and eighteenth centuries. There are those philosophers—Kant is the most obvious and seminal figure—who consider that we cannot, in fact, know anything about reality 'as it is in itself', so that ontology can be coherently conceived only as the science of our thought about being, rather than as the science of being as such. On the other hand, there are philosophers, many of whom would trace their allegiances back to Plato and Aristotle, who think that there is no obstacle in principle to our knowing at least something about reality as it is in itself. On behalf of this view, which I share, it may be urged that to deny the possibility of such knowledge is ultimately incoherent and self-defeating. The easiest way to sustain this charge is to point out that if, indeed, we could know nothing about reality as it is in itself, then for that very reason we could know nothing about our own thoughts about, or portrayals of, reality: for those thoughts or portrayals are nothing if not parts of reality themselves. In short, ontological questions—understood as questions about being rather than just about our thoughts about being—arise with regard to the ontological status of our thoughts, and of ourselves as thinkers of those thoughts: so that to attempt to recast all ontological questions as questions about our thoughts about what exists is to engender a regress which is clearly vicious.

This still leaves unanswered the question of how we can attain knowledge of being, or of reality 'as it is in itself', especially if ontology is conceived to be not an empirical but an *a priori* science. The answer that I favour divides the task of ontology into two parts, one which is wholly *a priori* and another which admits empirical elements. The *a priori* part is devoted to exploring the realm of metaphysical possibility, seeking to establish what kinds of things could exist and, more importantly, *co-exist* to make up a single possible world. The empirically conditioned part seeks to establish, on the basis of empirical evidence and informed by our most successful scientific theories, what kinds of things do exist in this, the actual

---

[1] See further my *The Possibility of Metaphysics: Substance, Identity, and Time* (Oxford: Clarendon Press, 1998), ch. 1, and my *A Survey of Metaphysics* (Oxford: Clarendon Press, 2002), ch. 1, where many of the points made in the present section of this chapter receive a fuller treatment.

world. But the two tasks are not independent: in particular, the second task depends upon the first. We are in no position to be able to judge what kinds of things actually *do* exist, even in the light of the most scientifically well-informed experience, unless we can effectively determine what kinds of things *could* exist, because empirical evidence can only be evidence for the existence of something whose existence is antecedently possible.

This way of looking at ontological knowledge and its possibility demands that we accept, whether we like it or not, that such knowledge is fallible—not only our knowledge of what actually does exist, but also our knowledge of what could exist. In this respect, however, ontology is nowise different from any other intellectual discipline, including mathematics and logic. Indeed, it is arguable that it was the mistaken pursuit of certainty in metaphysics that led Kant and other philosophers in his tradition to abandon the conception of ontology as the science of being for a misconception of it as the science of our thought about being, the illusion being that we can attain a degree of certainty concerning the contents of our own thoughts which eludes us entirely concerning the true nature of reality 'as it is in itself'.

1.2 ONTOLOGICAL CATEGORIES

I have described ontology as being concerned, in its *a priori* part, with what kinds of things can exist and co-exist. By 'kinds' here I mean *categories*, a term which is inherited, of course, from Aristotle, who wrote a treatise going under that title.[2] (Later I shall be using the term 'kinds' in a more restricted sense, to denote one ontological category amongst others, so it is important that no confusion should arise on this score.) And by 'things' I mean *entities*, that is, *beings*, in the most general sense of that term. Category theory, then, lies at the heart of ontology—but, properly understood, concerns categories conceived as categories of being, not, in Kantian style, as categories of thought. (There is, of course, also a branch of mathematics called 'category theory', but since ontology has the first claim on the term, I use it here without apology to the mathematicians concerned.)

Strangely, for much of the twentieth century, many philosophers, even those who were broadly sympathetic to the realist conception of ontology that I favour, saw no need for category theory to lie at the heart of metaphysics. This is because they imagined that all the purposes of ontology could be served, in effect, by set theory, perhaps in the belief that anything can be 'modelled' in set theory and that any adequate model can be substituted, without loss, for whatever it is supposed to be a model of.[3] Thus, for instance, they supposed that instead of talking about properties of objects, we could talk about sets of objects, or, more sophisticatedly, about functions from possible

---

[2] See Aristotle, *Categories and De Interpretatione*, trans. J. L. Ackrill (Oxford: Clarendon Press, 1963).
[3] For similar strictures, see Barry Smith, 'On Substances, Accidents and Universals: In Defence of a Constituent Ontology', *Philosophical Papers* 26 (1997), pp. 105–27, especially p. 107.

worlds (themselves conceived, perhaps, as sets of objects) to sets of objects 'at' or 'in' those worlds. For instance, the property of being red might be 'represented' as a function which has, for each possible world as an argument, the set of red objects in that world as the corresponding value. And functions themselves, of course, are also ultimately 'represented' as sets, namely, as sets of ordered pairs of their arguments and values (ordered pairs in turn being 'represented' as sets of sets in the standard Wiener–Kuratowski fashion).

Nothing could be more myopic and stultifying than this view that all the purposes of ontology can be served by set theory and set-theoretical constructions. Sets themselves comprise just one category of entities amongst many, and one which certainly could not be the sole category of entity existing in any possible world.[4] Even if we suppose that so-called 'pure' sets are possible—sets that have in their transitive closure only other sets, including the 'empty' set—there must be more kinds of thing in any possible world than just such sets. This is true even if it is also true that anything whatever can, in some sense, be 'modelled' set-theoretically. We should not conflate a model with what it is a model of. Indeed, there is a kind of unholy alliance between this way of doing ersatz ontology via set-theoretical constructions and the anti-realist conception of ontology as the science of our thoughts about, or representations of, reality. What is common to both approaches is the misbegotten conviction that we must and can substitute, without significant loss, models or representations of things for the things themselves.

So what, then, are ontological categories and which such categories should we acknowledge? How are such categories to be 'individuated', that is, identified and distinguished? Here I shall make two preliminary claims, neither of them expressed very precisely at this stage. First, ontological categories are hierarchically organized and, second, ontological categories are individuated by the distinctive existence and/or identity conditions of their members. The two claims are mutually dependent, furthermore. I have already mentioned some ontological categories in passing: for instance, the categories of *object*, *property* and *set*. A hierarchical relation is observable even here, since sets comprise a sub-category of objects: that is to say, a set is a special kind of object—namely, it is an abstract object whose existence and identity depend entirely upon the existence and identities of its members. And thus we see here too how the category of set is individuated in terms of the existence and identity conditions of the entities that belong to it. (I hasten to emphasize that the sense in which an entity 'belongs' to a category is not to be confused with the special set-theoretical sense in which something is a 'member' of a set: to indulge in this confusion would be to treat the categories themselves as sets, whereas in fact sets comprise just one ontological category amongst many. I should perhaps remark, indeed, that ontological categories are not themselves to be thought of as *entities* at all, nor, *a fortiori*, as comprising a

---

[4] See further my *The Possibility of Metaphysics*, ch. 12, and also my 'Metaphysical Nihilism and the Subtraction Argument', *Analysis* 62 (2002), pp. 62–73.

distinctive ontological category of their own, the category of *category*. To insist, as I do, that ontological categories are categories of being, not categories of thought, is not to imply that these categories are themselves *beings*.)

As a further illustration of the foregoing points, consider the following two sub-categories of object, each of which is a special kind of *concrete* object, in contrast with such abstract objects as sets and propositions: *masses*, or material bodies, on the one hand, and *living organisms* on the other. Entities belonging to these two categories have quite different existence and identity conditions, because a living organism, being the kind of thing that is by its very nature capable of undergoing growth and metabolic processes, can survive a change of its constituent matter in a way that a mere mass of matter cannot. A mere mass, being nothing but an aggregate of material particles, cannot survive the loss or exchange of any of those particles, any more than a set can undergo a change of its members. As a consequence, it is impossible to *identify* a living organism with the mass of matter that constitutes it at any given stage of its existence, for it is constituted by different masses at different stages.[5]

It is a matter of debate how, precisely, ontological categories are hierarchically organized, although the top-most category must obviously be the most general of all, that of *entity* or *being*. Everything whatever that does or could exist may be categorized as an 'entity'. According to one view, which I favour myself, at the second-highest level of categorization all entities are divisible into either *universals* or *particulars*.[6] A partial sketch of a categorial hierarchy embodying this idea and others that I have just outlined is provided in Fig. 1.1 below. I must emphasize its partial and provisional character. Other ontologists deny the very existence of universals, while yet others believe that all particulars are reducible to, or are wholly constituted by, coinstantiated or 'comprent' universals. Already here we see a kind of question that is central to ontology: a question concerning whether one ontological category is more 'fundamental' than another. Those ontologists who maintain that particulars are wholly constituted by coinstantiated universals are not denying—as some other ontologists do—the existence of either particulars or universals, but they are claiming that the category of universals is the more fundamental of the two. The point of such a claim is to effect an ontological economy. An ontologist who is never concerned to effect such economies is in danger of ending up with an ontological theory which amounts to nothing more than a big list of all the kinds of things that do or could exist: shoes and ships and sealing wax, cabbages and kings—not to forget dragons, witches, ectoplasm and the philosopher's stone.[7]

---

[5] See further my *Kinds of Being: A Study of Individuation, Identity and the Logic of Sortal Terms* (Oxford: Blackwell, 1989), ch. 7.

[6] For an alternative view, see Roderick M. Chisholm, *A Realistic Theory of Categories: An Essay in Ontology* (Cambridge: Cambridge University Press, 1996), or his 'The Basic Ontological Categories', in Kevin Mulligan (ed.), *Language, Truth and Ontology* (Dordrecht: Kluwer, 1992).

[7] Compare Frank Jackson, *From Metaphysics to Ethics: A Defence of Conceptual Analysis* (Oxford: Clarendon Press, 1998), pp. 4–5.

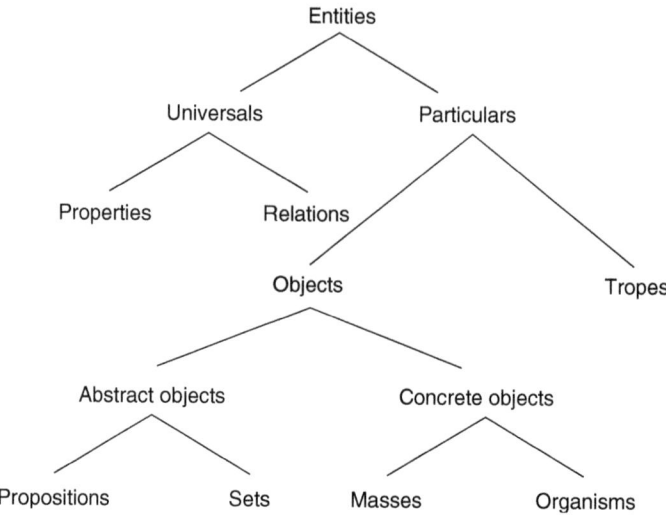

Figure 1.1. A fragment of the hierarchy of categories

1.3 SOME COMPETING ONTOLOGICAL SYSTEMS

This is where I can begin to make good my contention, implicit in the opening section of this chapter, that there have been recent advances in metaphysics. Progress has certainly been made of late in thinking about how ontological categories may be related to one another and, more especially, about which categories might have the strongest claim to being 'fundamental'. What does it mean to describe a certain ontological category as being 'fundamental'? Just this, I suggest: that the existence and identity conditions of entities belonging to that category cannot be exhaustively specified in terms of ontological dependency relations between those entities and entities belonging to other categories. This is why particulars cannot comprise a 'fundamental' ontological category if, in fact, they are wholly constituted by coinstantiated universals: for in that case, a particular exists just in case certain universals are coinstantiated and is differentiated from any other particular by the universals which constitute it. In point of fact, however, not many contemporary ontologists see much prospect in this account of particulars, not least because it implausibly excludes as metaphysically impossible a world in which two distinct particulars are qualitatively exactly alike—in other words, because it exalts to the status of a metaphysically necessary truth an implausibly strong version of Leibniz's principle of the identity of indiscernibles.[8]

---

[8] See further James Van Cleve, 'Three Versions of the Bundle Theory', *Philosophical Studies* 47 (1985), pp. 95–107.

I have been getting ahead of myself a little in talking of universals and particulars without offering any explicit account of the distinction between them. Even in this matter, however, there is controversy. Loosely, it is often said that universals are 'repeatable' and particulars 'non-repeatable' entities. By this account, the property of being red, or redness, conceived as a universal, is something that may be wholly and repeatedly present at many different times and places, whereas a particular red object is wholly confined to a unique space-time location and cannot 'recur' elsewhere and elsewhen.[9] There are problems with this way of characterizing the distinction between universals and particulars, but I shall not go into them here. Not surprisingly, however, a good many contemporary ontologists would like either to eliminate universals altogether from their inventories of existence or else to reduce them to particulars. This is the position of so-called trope theorists, for whom properties themselves are one and all particulars, with the redness of any one red object being numerically distinct from the redness of any other, even if the two objects in question resemble each other exactly in respect of their colours.

Another ontological distinction which requires some explication at this point is the distinction between *object* and *property*. Although for some ontologists this simply coincides with the distinction between particular and universal, clearly it does not for trope theorists. Objects are entities which possess, or 'bear', properties, whereas properties are entities that are possessed, or 'borne' by objects. Matters are complicated by the fact that properties themselves can, apparently, possess properties, that is, so-called 'higher-order properties'—as, for example, the property of being red, or redness, seemingly has the second-order property of being a colour-property. In view of this, one may wish to characterize an 'object' more precisely as being an entity which bears properties but which is not itself borne by anything else. This, however, is one traditional way of characterizing the category of *individual substance*—a way that may be found in some of the works of Aristotle, for instance.[10]

Trope theorists hold that objects, or individual substances, are reducible to tropes, that is, to properties conceived as particulars rather than as universals. On this view, an object, such as a certain individual flower, is wholly constituted by a number of 'compresent' tropes: it is, as it were, nothing over and above the particular properties that it possesses, such as a certain colour, shape, size, mass and so forth. It is, as they say, a 'bundle' of tropes, all of which exist in the same place at the same time. Trope theorists advertise as one of the main virtues of their theory the fact that it is a 'one-category' ontology—meaning by this that, according to their theory, there is only one fundamental ontological category, that of tropes. Objects, or individual substances, are regarded, as we have just seen, as being 'bundles' of tropes, depending for their existence and identity upon the tropes which constitute

---

[9] See, for example, David Armstrong, *Universals: An Opinionated Introduction* (Boulder, CO: Westview Press, 1989), pp. 98–9, and, for an objection, my *The Possibility of Metaphysics*, p. 156.
[10] For more on the category of substance, see Joshua Hoffman and Gary S. Rosenkrantz, *Substance Among Other Categories* (Cambridge: Cambridge University Press, 1994).

them, while universals, if they are wanted at all, are reducible to classes of resembling tropes—redness, thus, to the class of red tropes.

I may already have given the impression of such a diversity of opinion amongst contemporary ontologists as to undermine my own claim that advance has been made in modern metaphysics. But advance is not always made simply by arriving at a consensus of opinion. Sometimes it is made by the development of new theories and healthy argument between their adherents. This, indeed, is what very often happens in the empirical sciences too. However, it is time that I injected more order into my characterization of the rival ontological systems that are currently under debate.

To fix nomenclature, if only for the time being, let us operate with the terms *object*, *universal* and *trope*. An object is a property-bearing particular which is not itself borne by anything else: in traditional terms, it is an individual substance. A universal (at least, a first-order universal) is a property conceived as a 'repeatable' entity, that is, conceived as something that may be borne by many different particulars, at different times and places. And a trope is a property conceived as a particular, a 'non-repeatable' entity that cannot be borne by more than one object. Current ontological theories differ both over the question of the very existence of entities belonging to these three categories and over the question of which of the categories are fundamental. Of the many possible positions arising from different combinations of answers to these two questions, I shall pick out just four which have received some support in recent times.

First, then, there is the position of the pure trope theorists—such as Keith Campbell[11]—who regard tropes alone as comprising a fundamental category, reducing objects to bundles of compresent tropes and universals, if they are wanted at all, to classes of resembling tropes. A second position—espoused, for example, by David Armstrong[12]—acknowledges both objects and universals as comprising fundamental categories, while denying the existence of tropes. A third position—one that is currently championed by C. B. Martin[13]—acknowledges both objects and tropes as comprising fundamental categories, while denying the existence of universals or, again, reducing them to classes of resembling tropes. Unsurprisingly, the fourth position acknowledges all three categories of entity—object, universal and trope—as being fundamental, without denying, of course, that members of these categories stand in various ontologically significant relationships to one another. The distinguishing features of the four different ontological systems are set out in Table 1.1 below.

---

[11] See Keith Campbell, *Abstract Particulars* (Oxford: Blackwell, 1990). See also Peter Simons, 'Particulars in Particular Clothing: Three Trope Theories of Substance', *Philosophy and Phenomenological Research* 54 (1994), pp. 553–75.

[12] See David Armstrong, *A World of States of Affairs* (Cambridge: Cambridge University Press, 1997).

[13] See C. B. Martin, 'Substance Substantiated', *Australasian Journal of Philosophy* 58 (1980), pp. 3–10, and 'The Need for Ontology: Some Choices', *Philosophy* 68 (1993), pp. 505–22. See also C. B. Martin and John Heil, 'The Ontological Turn', *Midwest Studies in Philosophy* XXIII (1999), pp. 34–60.

Table 1.1. Four ontological systems

	Objects	Universals	Tropes
1	R	E/R	F
2	F	F	E
3	F	E/R	F
4	F	F	F

Key: F = Fundamental R = Reduced E = Eliminated

Before moving on, I want to make special mention of a variant of the fourth position which distinguishes between two different but equally fundamental categories of universals. This is the position that I favour myself, for reasons that I shall outline later. According to this position, there are two fundamental categories of particulars—objects and tropes—and two fundamental categories of universals: substantial universals, or *kinds*, whose particular instances are objects, and property-universals, whose particular instances are tropes. This is a position which some commentators have attributed to Aristotle on the basis of passages in his previously mentioned work, the *Categories*. It has also found some other adherents in modern times.[14]

1.4 STATES OF AFFAIRS AND THE TRUTHMAKER PRINCIPLE

At this point we need to reflect on some of the considerations that motivate current debate between the adherents of these different ontological systems. Of the four systems, perhaps the most popular today are pure trope theory on the one hand and the two-category ontology of objects and universals on the other. Pure trope theory is largely driven, it would seem, by a strong desire for ontological economy and a radically empiricist stance in epistemology, inspiring frequent appeals to Occam's razor and a nominalistic hostility to belief in the existence of universals. The ontology of objects and universals is motivated at least in part by the desire to provide an adequate metaphysical foundation for natural science, including most importantly laws of nature. Adherents of this ontological system typically hold that laws of nature can properly be understood only as consisting in relations between universals. But another important driving force in this case is commitment to the so-called *truthmaker principle*.[15] This is the principle that any true proposition or statement—or, at least, any contingently true proposition or

---

[14] See, for example, Barry Smith, 'On Substances, Accidents and Universals: In Defence of a Constituent Ontology', pp. 124–5.
[15] See David Armstrong, *A World of States of Affairs*, pp. 115 ff.

statement—must be *made* true by the existence of something appropriate in reality. (I set aside here the question of whether propositions or statements, or indeed sentences, are the primary bearers of truth and falsehood.)

It is a matter for some debate exactly what 'truthmaking' is, but on one plausible (if not entirely unproblematic) account of it, a truth-bearer is made true by a truthmaker in virtue of the truthmaker's existence entailing the truth of the truth-bearer. In the case of the contingent truth of a simple existential proposition, such as the proposition that Mars exists, it will then simply be a certain *object*—in this case, Mars itself—that is the truthmaker. But in the case of a contingently true predicative proposition, such as the proposition that Mars is red, the truthmaker, it seems, will have to be something in the nature of a *fact* or *state of affairs*—Mars's being red—which contains as constituents both an object, Mars, and a universal exemplified by that object, redness.[16] For the leading adherent of this sort of view, states of affairs are the building blocks of reality: the world is, in the words of David Armstrong, a world of states of affairs—recalling the famous opening remarks of Wittgenstein's *Tractatus*, 'The world is everything that is the case ... [it] is the totality of facts, not of things'.[17]

Saying that states of affairs are the building blocks of reality need not be seen as inconsistent with saying that the two fundamental ontological categories are the category of objects and the category of universals. On the view now under discussion, states of affairs are constituted by objects and universals, in the sense that these entities are the ultimate constituents of states of affairs. At the same time, it is held that objects and universals can only exist in combination with one another as constituents of states of affairs. Each category of entity may be conceived of as a distinct species of invariant across states of affairs. Objects recur in one way in different states of affairs, namely, as exemplifying different universals. And universals recur in another way in different states of affairs, namely, as being exemplified by different objects. Talk of objects 'recurring' in this sense is not at odds with their being particulars and so 'non-repeatable'. Their non-repeatability is a matter of their not being 'wholly present' at different times and places, in the way that universals supposedly are. As for states of affairs themselves, they are said to be particulars rather than universals, even though they contain universals as constituents: Armstrong speaks of this as 'the victory of particularity'.[18]

Not all ontologists who recognise the fundamental status of objects and universals are equally enamoured of states of affairs, however. They may have doubts about

---

[16] As Armstrong himself acknowledges, this claim may not seem compelling to believers in tropes, for at least some of whom Mars's particular redness suffices as a truthmaker of the proposition in question. See further, for example, Kevin Mulligan, Peter Simons and Barry Smith, 'Truth-Makers', *Philosophy and Phenomenological Research* 44 (1984), pp. 287–321 and Barry Smith, 'Truthmaker Realism', *Australasian Journal of Philosophy* 77 (1999), pp. 274–91. The latter paper also highlights some of the difficulties attending a simple entailment account of truthmaking.

[17] See David Armstrong, *A World of States of Affairs*, especially ch. 8. See also Ludwig Wittgenstein, *Tractatus Logico-Philosophicus*, trans. C. K. Ogden (London: Routledge and Kegan Paul, 1922).

[18] See David Armstrong, *A World of States of Affairs*, pp. 126–7.

the truthmaker principle or, at least, about the reification of states of affairs. There are certainly problems about treating facts or states of affairs as entities, let alone as the ultimate building blocks of reality. The existence and identity conditions of facts are hard to formulate in a trouble-free way. Perhaps the best-known problem in this connection is posed by the so-called 'Slingshot argument', which purports to reduce all facts to one fact, ironically called by Donald Davidson 'the Great Fact'.[19] The argument purports to show that, given certain allegedly plausible rules of inference, for any two true propositions P and Q, the expressions 'the fact that P' and 'the fact that Q' must have the same reference, if they refer to anything at all. The rules stipulate merely that in such an expression 'P' or 'Q' may be replaced, without the expression undergoing a change of reference, by any logically equivalent sentence or by any sentence in which a referring expression is replaced by another having the same reference. I shall not attempt to pass a verdict on the Slingshot argument here, but I do believe that it poses a significant challenge to the idea that states of affairs can be seen as the building blocks of reality, with objects and universals forming their 'constituents'.[20]

1.5 LAWS OF NATURE AND PROPERTIES AS WAYS OF BEING

I mentioned earlier the role that universals are thought by some ontologists to play in laws of nature. The issue here is whether laws can be seen as consisting in mere uniformities—or, as David Hume might have put it, 'constant conjunctions'—amongst particulars. For instance, does the law that planets move in elliptical orbits—Kepler's first law—simply amount to the fact that each and every individual planet moves in such an orbit? (I do not necessarily mean talk of a 'fact' here to carry any ontological weight: one may, if one is suspicious of facts, reconstrue what has just been said in terms of the truth of a proposition.) One apparent problem with such a suggestion is that not every individual planet does so move, because some—indeed, in reality, all—are subject to interference by the gravitational attraction of other bodies besides the star which they are orbiting.

More seriously still, the suggestion renders inexplicable our conviction that statements of natural law entail (or at least support) corresponding counterfactual conditionals. We want to say that if an actually planetless star had had a planet, then that planet would have moved in an elliptical orbit: but this cannot be entailed by the fact that each and every actually existing planet moves in an elliptical orbit. The answer to this problem, it is urged, is to say that the law consists in a relation

---

[19] See Donald Davidson, 'True to the Facts', in his *Inquiries into Truth and Interpretation* (Oxford: Clarendon Press, 1984). For wide-ranging discussion, see Stephen Neale, 'The Philosophical Significance of Gödel's Slingshot', *Mind* 104 (1995), pp. 761–825.
[20] See further my *The Possibility of Metaphysics*, pp. 241–3.

between two universals, the property of being a planet and the property of moving in an elliptical orbit—a relation of 'necessitation' which constrains any particular exemplifying the first property to exemplify the second as well.[21] For this constraint will apply not just in the actual world, but in any counterfactual situation—any possible world—in which those properties are related in the same way as they are in the actual world, and thus in any possible world in which the law in question obtains. The pure trope theorist, in denying universals, is apparently committed to a 'constant conjunction' conception of laws, as is the advocate of an ontology admitting only objects and tropes as fundamental entities.

In another respect, however, an advocate of the latter sort of ontology can to some extent find common cause with the advocate of universals on the matter of property-bearing. For the pure trope theorist, individual objects are just 'bundles' of 'compresent'—that is, spatiotemporally coinciding—tropes. However, this seems to grant to tropes a kind of ontological independence which they plausibly cannot have. It is not clear, on this view, why the tropes in any given bundle should not separate from one another and either float free of other tropes altogether or migrate to other trope-bundles. It has seemed better to many ontologists to conceive of properties—whether they be regarded as universals or as particulars—as *ways* objects are.[22] An object's redness, thus, is its way of being coloured and its roundness, say, is its way of being shaped. If one thinks that different objects may literally be coloured or shaped in the very *same* way—that is, in numerically the same way—then one is thinking of these 'ways' as universals. Otherwise, one is thinking of them as trope-like entities—particular 'ways', or, to revert to a more traditional terminology, *modes*. Opponents of pure trope theory will say that it makes no sense to suppose that an object—something that *has* properties such as redness and roundness—can just be constituted by those very properties, being nothing over and above the sum of its properties. To suppose this is, they will say, quite literally to make a 'category mistake'. It is to confuse an object's properties with its *parts*: for the parts of an object, if it has any, are themselves objects, with properties of their own.[23]

In reply, the trope theorist may challenge opponents to say what *more* there is or can be to an object than the properties that it bears. This is a dangerous question for the opponents of trope theory, for they may be tempted to say that objects do indeed possess an additional 'ingredient' or 'constituent', over and above the properties that they bear, characterizing this additional constituent as a 'substratum' or 'bare particular'—that is, an entity which is not itself a property, nor yet a propertied object, but a constituent of an object which plays the role of 'bearing'

---

[21] See David Armstrong, *What is a Law of Nature?* (Cambridge: Cambridge University Press, 1983), p. 85.
[22] See Jerrold Levinson, 'Properties and Related Entities', *Philosophy and Phenomenological Research* 39 (1978), pp. 1–22.
[23] See C. B. Martin, 'Substance Substantiated', for such a criticism. Other philosophers, however, contend that tropes are indeed parts of objects, but *dependent* rather than independent parts.

that object's properties.²⁴ In my view, those who go down this road make the mistake of conceding in the first place that an object's properties are 'constituents' of the object. For it was this move that committed them to finding some *further* 'constituent' of an object once they denied the trope theorist's contention that an object is *wholly* constituted by its properties. The proper thing to do, I suggest, is just to emphasize again that an object's properties are *ways* it is and say that the object itself is the 'bearer' of its properties, not some mythical 'constituent' of the object that is somehow buried within it and inescapably hidden from view.

Suppose we accept that universals must be included in our ontology as fundamental in order to account for the ontological status of natural laws and accept too that individual objects comprise a fundamental category of entities, irreducible to their properties, whether the latter are conceived as universals or as particulars. What is to be said for including properties *both* as universals *and* as particulars in our ontology? Mainly this, I think: it seems that only particulars can participate in causal relationships and that an object participates in such relationships in different ways according to its different properties. Thus, it is a rock's *mass* that explains the depth of the depression it makes upon falling on to soft earth, whereas it is the rock's *shape* that explains the shape of the depression. Perception itself involves a causal relationship between the perceiver and the object perceived and we perceive an object by perceiving at least some of its properties—we perceive, for instance, a flower's colour and smell. But this seems to require that what we thus perceive are items that are unique to the object in question—*this* flower's redness and sweetness, say, as opposed to a universal redness and sweetness that are also exemplified by other, exactly resembling flowers.²⁵ For, surely, in seeing and smelling this flower, I cannot be said to perceive the colour and smell of any other flower.

The only response to this last point that seems available to the opponent of properties conceived as particulars is to say that what I see and smell in such a case is not, literally, the redness and sweetness of the flower as such, these allegedly being universals, but, rather, the *fact* that the flower is red and the *fact* that it is sweet, these facts being construed as particulars which enter into causal relations when perception occurs. But this then saddles us again with an ontology of facts or states of affairs, which we have seen to be open to objection.

1.6 THE FOUR-CATEGORY ONTOLOGY

If the foregoing diagnosis is correct, we should gravitate towards the fourth system of ontology identified earlier, the system which acknowledges three distinct ontological categories as being fundamental and indispensable—the category of *objects*, or individual substances; the category of *universals*; and the category of

---

²⁴ I criticize this view in my 'Locke, Martin and Substance', *Philosophical Quarterly* 50 (2000), pp. 499–514.   ²⁵ See further my *The Possibility of Metaphysics*, p. 205.

tropes, or, as I shall henceforth prefer to call them, *modes*. It is then but a short step to my preferred variant of this system, which distinguishes between two fundamental categories of universal, one whose instances are objects and the other whose instances are modes. This distinction is mirrored in language by the distinction between *sortal* and *adjectival* general terms—that is, between such general terms as 'planet' and 'flower' on the one hand and such general terms as 'red' and 'round' on the other.[26] The former denote *kinds* of object, while the latter denote *properties* of objects. Individual objects are particular instances of kinds, while the modes of individual objects are particular instances of properties. If a distinctive term is wanted to speak of properties thus conceived as universals, the term *attribute* will serve, although in what follows I shall for the most part either allow context to eliminate any ambiguity or else speak explicitly of property-universals. I believe that this system of ontology has a number of advantages over all of its rivals, a few of which I shall briefly sketch now.

The four-category ontology—as I like to call it—provides, I believe, a uniquely satisfactory metaphysical foundation for natural science.[27] It can, for instance, account for the ontological status of natural laws by regarding them as involving universals, but not simply property-universals. Rather, laws typically involve both kinds and either properties or relations. Take, for example, the law that I expressed earlier by means of the law-statement 'Planets move in elliptical orbits'. According to the most popular current view of laws as involving universals—the view championed by David Armstrong—this law consists in the fact that a second-order relation of necessitation obtains between the first-order properties of being a planet and moving in an elliptical orbit. I say, rather, that the law consists in the fact that the property of moving in an elliptical orbit characterizes the kind *planet*. In this way, I both obviate the need to appeal to any second-order relation and provide an account of the ontological status of laws which more closely reflects the syntactical structure of law-statements. For, as I have pointed out elsewhere, the standard form of law-statements in natural language is that of dispositional predications with natural kind terms in subject-position, other examples being 'Gold is fusible', 'Electrons are negatively charged' and 'Mammals are warm-blooded'.[28] Notice, in this connection, that the predicate in 'Planets move in elliptical orbits' is clearly dispositional in force: the law-statement is an expression of how planets are *disposed* to move, under the gravitational influence of a star. And this, indeed, is why such a law-statement is not falsified by the fact that the *actual* movements of planets often deviate from the elliptical orbits in which they would move if they were not subject to interference by the gravitational forces exerted by other planets. I should add that some laws are genuinely relational, such as the law that electrons and protons attract one another: but here the relation is not one in which only

---

[26] See further my *Kinds of Being*, ch. 2.
[27] I first introduced this name for the present ontological system in my *The Possibility of Metaphysics*, pp. 203–4. [28] See my *Kinds of Being*, ch. 8.

universals can stand to one another, so it is not in that sense a 'second-order' relation, like the relation of 'necessitation' invoked by the rival universalist account of laws.

Next, the four-category ontology can account for the distinction between dispositional and occurrent (or 'categorical') states of objects—between, for instance, an object's being fusible and its actually melting, or between an object's being soluble and its actually dissolving. Various other accounts of this distinction have been offered recently by metaphysicians, none of which, in my view, is entirely satisfactory. Attempts to analyse disposition statements in terms of counterfactual conditionals all founder on the fact that the manifestation of a disposition can always be inhibited or prevented by interfering factors.[29] Thus, for example, '*O* is water-soluble' cannot be analysed as 'If *O* were immersed in water, then *O* would be dissolving', nor can the antecedent of this counterfactual be expanded by any finite list of specifiable additional conditions in a way which will secure its logical equivalence with the original disposition statement. Merely adding the catch-all *ceteris paribus* condition that 'all other things are equal', or 'nothing interferes', simply serves to trivialize the proposed analysis.

According to the four-category ontology, the distinction between dispositional and occurrent states of objects may be explained in the following way. An object possesses a *disposition* to *F* just in case it instantiates a kind which is characterized by the property of being *F*. Thus, for example, an object *O* has a disposition to be dissolved by water just in case *O* instantiates a kind, *K*, such that the law obtains that water dissolves *K*. Here, *K* might be, say, the kind *sodium chloride* and the law, correspondingly, the law that water dissolves sodium chloride. As we have already seen, by my account of laws, laws themselves are dispositional in force. And, indeed, this is borne out in the present case by the fact that the law just stated can be equally well expressed by the sentence 'Sodium chloride is water-soluble'. On the other hand, an object is *occurrently F* just in case it possesses a mode which is an instance of the property of being *F*, that is, a mode of the universal *F*ness. To apply this sort of analysis to the case of an object *O*'s occurrently being dissolved by some water, we merely need to invoke *relational* modes, whereupon we can analyse this occurrent state as obtaining just in case *O* and some water are related by a mode which is an instance of the universal relation of dissolution. By the account of laws which I favour, it is, of course, the fact that this same universal relation holds between the kinds *water* and *sodium chloride* that constitutes the law that water dissolves sodium chloride. Thus it emerges that the distinction between the dispositional and the occurrent simply reflects, ultimately, the ontological distinction between the domain of universals and the domain of particulars.

Combining this observation with my earlier remarks about perception, we can now understand why it is that an object's occurrent states are perceptible but its dispositions are not. For what we can perceive of an object are its modes—its *particular* 'ways of being'—and it is in virtue of possessing these that the object is

---

[29] See C. B. Martin, 'Dispositions and Conditionals', *Philosophical Quarterly* 44 (1994), pp. 1–8.

in various occurrent states, say of melting or dissolving. By contrast, the object is in various dispositional states in virtue of instantiating kinds which are characterized by various property-universals, that is, kinds which are subject to various laws—and this is not the sort of circumstance that perception can acquaint us with directly (although, of course, it can provide empirical evidence for it).

The four-category ontology has no difficulty in saying what 'ties together' the particular properties—that is, the *modes*—of an object. An object's modes are simply 'particular ways it is': they are characteristics, or features, or aspects of the object, rather than constituents of it. If properties were constituents of an object, they would need, no doubt, to be tied together somehow, either very loosely by coexisting in the same place at the same time, or more tightly by depending in some mysterious way either upon each other or upon some still more mysterious 'substratum', conceived as a further constituent of the object, distinct from any of its properties. It is precisely because a mode is a particular way this or that particular object is that modes cannot 'float free' or 'migrate' from one object to another—circumstances that pure trope theorists seem obliged to countenance as being at least metaphysically possible. Moreover, the four-category ontology allows us to say that the properties of a *kind* are tied to it, in the laws to which it is subject, in a manner which entirely parallels, at the level of universals, the way in which an individual object's modes are tied to that object. In both cases, the tie is simply a matter of the 'characterization' of a propertied entity by its various properties and consists in the fact that the properties are 'ways' the propertied entity is.

Fig. 1.2 below may help to highlight the main structural features of the four-category ontology as I have just outlined it. I shall return to it (or slightly modified versions of it) frequently in later chapters of this book, where—in deference to

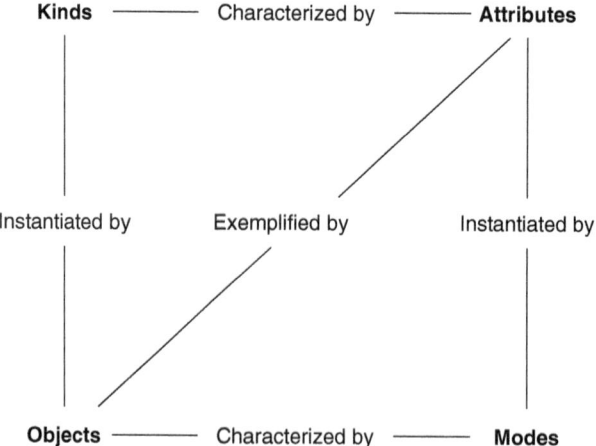

Figure 1.2. The four-category ontology

tradition—I shall call it 'the Ontological Square'. In this diagram I use the term 'attribute', as suggested earlier, to denote the category of property-universals and, for simplicity of presentation, I am ignoring relational universals. An object O may exemplify an attribute A in either of two ways. O may instantiate a kind K which is characterized by A, in which case O exemplifies A *dispositionally*. Alternatively, O may be characterized by a mode M which instantiates A, in which case O exemplifies A *occurrently*.

It may perhaps be doubted whether the four-category ontology provides an adequate metaphysical foundation for the more esoteric reaches of modern physics, such as the general theory of relativity and quantum physics. But I believe that even here it will serve well enough. The examples of 'objects', 'kinds', 'attributes' and 'modes' that I have so far utilized have been for the most part fairly familiar and mundane ones. But nothing hinders us from saying, if need be, that relativistic space-time has the status of an individual substance or object, with the consequence, perhaps, that the entities that we are ordinarily apt to regard as objects—such as material bodies—are 'really' just spatiotemporally continuous successions of space-time modes. This is a view of the material world which, indeed, is prefigured in the metaphysical system of Spinoza. Again, we need not take a stand on the issue of whether the ontology of quantum physics is best construed in a way which treats quantum entities as particles—a kind of object—or as modes of a quantized field. Either way, the four-category ontology will admit of application.

It is important to stress, then, that metaphysics should not be in the business of dictating to empirical scientists precisely how they should categorize the theoretical entities whose existence they postulate. Metaphysics supplies the categories, but how best to apply them in the construction of specific scientific theories is a matter best left to the theorists themselves, provided that they respect the constraints which the categorial framework imposes. So long as the empirical sciences invoke laws for explanatory purposes and appeal to perception for empirical evidence, the four-category ontology will, I believe, adequately serve as a metaphysical framework for the scientific enterprise. That *some* metaphysical framework is necessary for the success of that enterprise and that its formulation is not the business of any special science, but only that of the general science of being, or ontology, I hope to be by now beyond dispute.

2

The Four-Category Ontology and its Rivals

2.1 CENTRAL PRINCIPLES OF THE FOUR-CATEGORY ONTOLOGY

There are many ways of constructing a system of ontological categories and little agreement as to what, exactly, we should understand by the expression 'ontological category'. As I understand this expression, an ontological category is a *kind of being*, that is, a kind of entities, membership of which is determined by certain distinctive existence and identity conditions whose nature is determinable *a priori*. Such a kind, then, is not to be confused with so-called 'natural' kinds, referred to by specific sortal terms such as 'tiger' or 'gold'. For, although the members of such natural kinds will, of course, be entities belonging to appropriate ontological categories—as, for example, a tiger is a *living organism* and a portion of gold is a *quantity of matter or stuff*—the nature of such kinds is determinable only *a posteriori*, by scientific observation and experimentation. Ontological categorization, as I understand it, operates at a higher level of abstraction than does scientific taxonomizing, and the latter presupposes the former.[1] But some ontological categories are more basic than others, so that such categories can be organized into a hierarchy, perhaps in more ways than one. Thus, the category of *living organism*, to which an individual tiger belongs, is a sub-category of the higher-level category of *individual substance* or *particular 'object'* (in one sense of that dangerously ambiguous term). The basic categories of an ontological system are those occupying the highest level and are at the same time those by reference to which the existence and identity conditions definitive of the lower-level categories are specified. For instance, what makes living organisms different, in respect of their existence and identity conditions, from certain other categories of individual substance, is that they may survive a systematic change of their constituent matter, provided that they continue to exemplify a specific form appropriate to their natural kind.

---

[1] See further my *The Possibility of Metaphysics: Substance, Identity, and Time* (Oxford: Clarendon Press, 1998), ch. 8.

My concern in this chapter is to defend a certain position regarding what *basic* ontological categories we should recognize. The position in question seems to be implicit in what is perhaps the first treatise ever devoted explicitly and wholly to our subject, Aristotle's *Categories*.[2] It is that the basic categories are four in number, two of them being categories of particular and two being corresponding categories of universal. The terms 'particular' and 'universal' themselves, we may say, do not strictly denote categories, however, because they are *transcategorial*, applying as they do to entities belonging to different basic categories.

The first category and in a certain sense the most fundamental—even though in another sense all four of our categories are equally 'basic'—is the category of *individual substance* or *particular 'object'*, which I have already mentioned. Corresponding to this category of particular, there is a basic category of universal, namely, the category of *substantial universal* or *substantial kind*, the correspondence consisting in the fact that each individual substance necessary instantiates—is a particular instance of—some substantial kind. The natural kinds *tiger* and *gold*, cited earlier, are examples of substantial universals or, as I shall henceforth briefly refer to them, *kinds* (on the understanding, of course, that by such a 'kind' I do not mean an ontological category, for reasons already given). The other basic category of universal is the category of *properties and relations*, or (as I also call them) *non-substantial universals*, examples of these being the properties redness and squareness and the spatial relation of betweenness. Obviously, the distinction between the two basic categories of universal is reflected in language in the distinction between substantival general terms on the one hand and adjectival and relational general terms on the other. This leaves one further basic category, which is the category of particular corresponding to the category of non-substantial universals—the category whose members are particular instances of properties and relations, that is, *property- and relation-instances* or, as they are sometimes called, *tropes*.[3] My own preferred term for (monadic) property-instances is *modes*, though they have also been variously called individual accidents, particular qualities and abstract particulars.

We need to be clear about exactly how items in these four categories are related to one another. I have already remarked that the relationship between an individual substance and its kind is one of *instantiation*, as is the relationship between a property- or relation-instance—a trope—and the corresponding non-substantial universal.[4] A particular tiger is an instance of the kind *tiger*, and a particular redness—say of a certain individual flower—is an instance of the non-substantial universal or property *redness*. But this still leaves certain other crucial relationships between members of the different basic categories undescribed. Most importantly,

---

[2] See the *Categories*, ch. 2, in *Aristotle's Categories and De Interpretatione*, trans. J. L. Ackrill (Oxford: Clarendon Press, 1963).
[3] See further Keith Campbell, *Abstract Particulars* (Oxford: Blackwell, 1990).
[4] For more on instantiation, see my *Kinds of Being: A Study of Individuation, Identity and the Logic of Sortal Terms* (Oxford: Blackwell, 1989), ch. 3.

there is the relationship between a property- or relation-instance and the individual substance or substances to which that instance belongs or which that instance relates. I call this the relationship of *characterization*. A particular redness *characterizes* the individual substance whose redness it is and a particular betweenness *characterizes* the three individual substances (taken in a certain order) which it relates. (Here I leave aside the question of whether points of space exist as relata of betweenness relations and, if so, whether they qualify as 'individual substances'.) Paralleling this relationship at the level of particulars is a corresponding relationship at the level of universals. For, just as we may say that a particular redness characterizes a certain individual substance, such as a particular flower, so we may say that the property or non-substantial universal *redness* characterizes a certain substantial universal or kind, such as the natural kind *tomato*. Speaking quite generally, then, and prescinding from the distinction between universals and particulars, we may say that characterization is a relationship between property- or relation-like entities on the one hand and substantial entities on the other. We may summarize our proposals so far in the form of the diagram in Fig. 2.1 which—as I mentioned in the previous chapter—I like to call 'the Ontological Square'.

It will be clear from this diagram that one important species of relationship between entities of different basic categories has yet to be given a name by us. This is the relationship between an individual substance and some non-substantial universal to which it is indirectly related either via one of its property- or relation-instances or via its substantial kind. I propose to call this sort of relationship *exemplification*. Thus, in the present system of ontology, it is vital to distinguish clearly between instantiation, characterization, and exemplification. An individual ripe tomato *instantiates* the kind *tomato*, is *characterized* by a particular redness, and *exemplifies* the non-substantial universal or property *redness*. The tomato's

Figure 2.1. The Ontological Square (I)

particular redness, by contrast, *instantiates* the property *redness*. And the kind *tomato* is *characterized* by the property *redness*.

2.2 PROPERTIES AS UNIVERSALS AND PROPERTIES AS PARTICULARS

The ontological system that I have been describing will certainly strike some metaphysicians as being unnecessarily complex, with many redundant features. Do we really need four basic ontological categories and (at least) three different kinds of relationship between members of those categories—instantiation, characterization, and exemplification? Some philosophers, for example, while happy to countenance the existence of universals, may think it extravagant to include in our ontology property- and relation-instances, in our sense, in addition to both individual substances and non-substantial universals.[5] Why not say, as some philosophers do, that non-substantial universals such as *redness* are instantiated by individual substances, that is, that things such as an individual tomato or an individual flower are the particular instances of such a universal, rather than the supposed particular 'rednesses' of such individual substances? From our point of view, this is to obliterate the distinction between instantiation and exemplification, but some may say that this is a welcome economy.

I have various responses to philosophers of this persuasion. For one, consider the fact that we can, it seems clear, *perceive* at least some of the properties of individual substances, such as an individual tomato's redness. But perception, it would seem, is necessarily of particulars, since only particulars can enter into causal relations or literally possess causal powers—and perception necessarily involves a causal relation between the perceiver and what is perceived. Moreover, not only can we perceive at least some of the properties of individual substances, we can also perceive them undergo *changes* in their properties. When I see a leaf change from green to brown as it is burnt by a flame, I seem to see its former greenness go out of existence and its new brownness come into existence. But neither greenness nor brownness the *universals* are affected by such changes: they both exist before and after the leaf is burnt. So, it seems, it must be the leaf's *particular* greenness and brownness that I see ceasing and beginning to exist respectively. To this it may be replied that what I see is not a particular greenness ceasing to exist and a particular brownness beginning to exist, but just greenness and brownness—the universals—beginning and ceasing to be *exemplified* by the leaf. However, the change that I see is *in the leaf itself*, not in its relationship to something universal that is unchanging. Furthermore, we may ask *in virtue of what* it is that the leaf now exemplifies brownness where it formerly exemplified greenness. Surely, it is in virtue of how the leaf

---

[5] See, for example, D. M. Armstrong, *A World of States of Affairs* (Cambridge: Cambridge University Press, 1997).

now is, in all its particularity—that is, it is in virtue of the *particular* characteristics of the leaf, one of which is its particular colour, that is, its particular brownness.

Those who believe in properties as universals only, not as particulars, are apt to say that universals are 'wholly present' in the various individual substances which exemplify them—for instance, that the very same universal redness is 'wholly present' in two different tomatoes which exactly match one another in hue.[6] And by this they mean to imply that a universal can be wholly—that is, *all* of it can be—in two different places at the same time. They tell us not to worry that this seems to make no sense, assuring us that we have this impression only because we are mistakenly tempted to assimilate the spatiotemporal location of universals to that of particulars, which, indeed, cannot be wholly in two different places at once. But, none the less, I am far from convinced that it does make sense to say that *anything*, be it universal or particular, can be wholly in two different places at once. This is because the relation of *being wholly in the same place as* appears to be an equivalence relation and therefore a symmetrical and transitive relation, which poses the following difficulty. Suppose that tomatoes *A* and *B* exemplify exactly the same shade of redness and that this universal is both wholly in the same place as *A* and wholly in the same place as *B*. Then it seems to follow, given the symmetry and transitivity of the relation *being wholly in the same place as*, that tomato *A* is wholly in the same place as tomato *B*—which we know to be necessarily false, given the non-identity of *A* and *B*.

To this it may be replied, perhaps, that the relation *being wholly in the same place as* is not in fact a symmetrical relation, so the fact that the universal in question is wholly in the same place as tomato *A* does not imply that tomato *A* is wholly in the same place as the universal. Indeed, it may be urged that tomato *A* plainly is *not* wholly in the same place as the universal, because the universal, unlike tomato *A*, is also wholly in the same place as tomato *B*. But this response strikes me as being both unprincipled and question-begging. Certainly, according to the theorists whose view is under scrutiny, *all* of the tomato is in a place where *all* of the universal is located. (I ignore here the quite separate issue of whether persisting things like tomatoes should be deemed to have 'temporal' parts.) If this doesn't mean that tomato *A* is wholly in the same place as the universal—that they are wholly co-located—then it is altogether obscure to me what 'being wholly in the same place as' can possibly mean. But if tomato *A* is wholly in the same place as the universal and the universal is wholly in the same place as tomato *B*, then it seems to me to follow, ineluctably, that tomato *A* is wholly in the same place as tomato *B*—and this is absurd.

At the very least, I think we can say that those who speak of universals being 'wholly present' in many different places at once owe us a much more perspicuous account of what they could possibly mean by saying this. My own suspicion is that what such theorists are doing, when they say this, is precisely conflating a

---

[6] See, for example, D. M. Armstrong, *Universals: An Opinionated Introduction* (Boulder, CO: Westview Press, 1989), pp. 98–9.

non-substantial universal with its particular property-instances. For what are in many different places at once, I should say, are the various particular property-instances, although these are united by the fact that they are all instances of exactly the same universal. But the universal itself cannot, I think, properly be said to have a location at all, much less many locations. This, however, does not make it some queer sort of 'Platonic' entity, inhabiting an 'ideal' realm somehow isolated from the world of things in space and time. The universal doesn't have to exist 'elsewhere', just because it doesn't have a location in space: it just has to *exist*, but without any spatial determination to its manner of existing. We can still say, indeed, that its manner of existing is, in a perfectly good sense, 'immanent' (rather than 'transcendent'), inasmuch as it exists only 'in' or 'through' its particular instances, precisely insofar as they instantiate it.[7] We can insist, thus, that there can be no *uninstantiated* universals and that particulars enjoy a kind of ontological priority over universals, just as Aristotle believed.

2.3 SUBSTANTIAL UNIVERSALS AND SUBSTANTIAL PARTICULARS

So far, I have been defending the four-category ontology against those who favour including universals, but not property-instances, in our ontology. But, of course, there are other philosophers who are happy to include property-instances but are opposed to including universals. Against these philosophers, I have another set of objections. But first I should defend my view that universals fall into two distinct basic categories, the substantial and the non-substantial, which many philosophers may dismiss as being overly reliant upon a superficial grammatical distinction. In point of fact, I do not at all think that metaphysics should be conducted entirely through the filter of language, as though syntax and semantics were our only guides in matters metaphysical—although it should hardly be surprising if natural language does reflect in its structure certain structural features of the reality which it has evolved to express.[8] The distinction between the substantial and the non-substantial must, however, be simultaneously defended at two different levels—at the level of particulars and at the level of universals. On the one hand, we must defend the ontological status of individual substances as basic entities against those who would represent them as being, or reduce them to, 'bundles' of 'compresent' property-instances or tropes.[9] On the other hand, we

---

[7] See further my 'Abstraction, Properties, and Immanent Realism', in Tom Rockmore (ed.), *Proceedings of the Twentieth World Congress of Philosophy, Volume II: Metaphysics* (Bowling Green, OH: Philosophy Documentation Center, 1999), pp. 195–205.

[8] See further my *The Possibility of Metaphysics*, ch. 1.

[9] See, for example, Campbell, *Abstract Particulars*, and Peter Simons, 'Particulars in Particular Clothing: Three Trope Theories of Substance', *Philosophy and Phenomenological Research* 54 (1994), pp. 553–75.

must defend the ontological irreducibility of substantial kinds against those who would reduce them to complexes of co-exemplified non-substantial universals.[10] Let me turn to this second strand of the project first.

It cannot be denied that there is an intimate relationship between a substantial universal, such as the natural kind *gold*, and certain non-substantial universals that are typically exemplified by individual substances of that kind—a golden colour, a certain melting point, a certain density, malleability, ductility, solubility in *aqua regia*, and so forth. But we plainly cannot say that an individual substance necessarily instantiates the kind *gold* if and only if it exemplifies all of these typical properties, for any substantial universal like this may have untypical individual exemplars.[11] It may also be that gold has a 'scientific essence', consisting in its being constituted by atoms possessing an atomic number of 79, in which case the predicates 'is gold' and 'is constituted by atoms possessing an atomic number of 79' would appear to be, of metaphysical necessity, co-extensive. However, 'is gold', in this context, means 'is made of gold', and 'gold' as a substantival general term must be distinguished from the predicative expression 'made of gold', or 'golden'. The substantival general term refers to a kind of stuff, which as a matter of natural law is characterized by many non-substantial universals, such as those cited a moment ago. But there is, in principle, no finite limit to the number of such characteristics nomically tied to the nature of gold: 'gold', used as a substantival general term, is not a way of denominating the totality of such characteristics, but rather a way of referring to what it is—a kind of substance—that *bears* those characteristics, as a matter of natural necessity. However, a fuller defence of this position will only become possible when, in due course, I say more about the nature of natural laws and natural necessity.

At the level of particulars, as opposed to that of universals, a defence of the distinction between the substantial and the non-substantial means defending the ontological irreducibility of the category of individual substance both against those who, as mentioned earlier, try to reduce individual substances to bundles of compresent tropes and against those who, even more implausibly, try to reduce them to bundles of co-exemplified universals. The latter strategy is made vulnerable by its apparent commitment to an implausibly strong version of Leibniz's principle of the identity of indiscernibles, so I shall ignore it from now on. The former strategy is not subject to this objection, but is still, I believe, fatally flawed, its basic problem being that it can provide no adequate account of the existence and identity conditions of tropes or, as we have been calling them, property-instances.[12]

---

[10] See, for example, Armstrong, *A World of States of Affairs*, pp. 65–8.

[11] On this point and others made in this paragraph of the present chapter, see further my *Kinds of Being*, ch. 8.

[12] See further my *The Possibility of Metaphysics*, pp. 205 ff. It should be noted that the claim that the so-called Bundle Theory is committed to the identity of indiscernibles has recently been challenged: see Gonzalo Rodriguez-Pereyra, 'The Bundle Theory is Compatible with Distinct but Indiscernible Particulars', *Analysis* 64 (2004), pp. 72–81.

Property-instances are ontologically dependent entities, depending for their existence and identity upon the individual substances which they characterize, or to which they 'belong'. A particular redness or squareness can, ultimately, be identified as the particular property-instance that it is only by reference to the individual substance which it characterizes. This is not an epistemic point but a metaphysical one: it concerns individuation in the metaphysical rather than in the cognitive sense—that is, individuation as a determination relation between entities rather than individuation as a kind of cognitive achievement. And this is the reason why it makes no sense to suppose that particular property-instances could exist free-floating and unattached to any individual substance or migrate from one individual substance to another.

Of course, one reason why some philosophers believe that what we have been calling 'individual substances' are reducible to, or consist in, bundles of compresent tropes, is that they think—erroneously, as I consider—that to deny this commits one to an indefensible doctrine of 'bare particularity' or to the notion of a propertyless 'substratum' that somehow 'supports' and 'unites' the properties of a single object. Their mistake is to think that we can only deny that an individual substance, or particular 'object', is the sum of its particular properties or tropes by contending, instead, that it somehow consists of these properties *plus* an extra non-qualitative ingredient, its 'substratum' or 'bare particular'. Not so. The underlying error is to think that an individual substance is any kind of *complex* at all, at least insofar as it is propertied in various ways. Individual substances may be complex in the sense of being *composite*, that is, in the sense of being composed of lesser substantial parts, as, for example, a living organism is composed of individual cells. But an individual substance's (particular) *properties*—its modes, as I prefer to call them—are not items of which it is *composed*.

There is, I think, an implicit conflation between parts and properties in the trope-bundle view of objects.[13] An individual substance has many properties—perhaps even infinitely many (if we include relational properties)—and in this sense is multifaceted: but this does not amount to any sort of ontological complexity in its constitution. *Of course* an individual substance's properties need no 'substratum' to 'support' them, if by such a substratum we mean some entity which somehow lacks properties 'of its own' and is some sort of non-qualitative ingredient of the individual substance. But the properties do indeed need 'support', in the sense that they are ontologically dependent entities which can only exist as the properties of that very individual substance. However, it is the individual substance *itself* which provides their 'support' in this entirely

---

[13] Compare C. B. Martin, 'Substance Substantiated', *Australasian Journal of Philosophy* 58 (1980), pp. 3–10. Martin has a two-category ontology of particular properties and particular substances. I endorse his recognition of this irreducible distinction, but disagree with some aspects of his conception of particular substances: see my 'Locke, Martin and Substance', *Philosophical Quarterly* 50 (2000), pp. 499–514.

legitimate sense, and this it can do without the spurious aid of some mysterious 'substratum'.[14] There is no mystery here as to *how* individual substances can perform this 'supporting' role, for once we recognize the category of individual substance as basic and irreducible and the category of property-instance as correlative with it, we can see that their having such a role is part of their essential nature. Explanation—even metaphysical explanation—must reach bedrock somewhere, and this, according to the four-category ontology, is one place where bedrock is reached. The idea that some more fundamental explanation is somehow available, if only we can probe reality more deeply, is, I think, just an illusion born of some of the confusions mentioned above.

2.4 IN DEFENCE OF UNIVERSALS

The stage which my defence of the four-category ontology has now reached is this. I have defended the inclusion of property-instances in our ontology. I have also defended the distinction between the substantial and the non-substantial, both at the level of universals and at the level of particulars. This means that I have thereby defended the inclusion of individual substances in our ontology as forming a basic and irreducible category of particulars additional to, and correlative with, the category of property-instances (in which I also include, by implication, relation-instances). But although I have defended the distinction between substantial and non-substantial universals against those supporters of universals who would reduce these two categories to one, I have said nothing yet explicitly to defend the four-category ontology against those who deny the existence of universals altogether. Here my main argument makes common cause with certain other defenders of universals who differ from me in not recognizing the distinction between substantial and non-substantial universals as fundamental: but I believe that recognizing that distinction helps to strengthen this argument against the opponents of universals in some important ways. I should say at once that I do not have in mind here any species of 'one over many' argument, to the effect that varieties of particularism such as resemblance nominalism cannot adequately account for the meanings that we attach to general terms or our ability to classify particulars—arguments which appeal at least partly to semantic or psychological considerations rather than to purely metaphysical ones.

The argument that I have in mind is one which contends that the ontological status of natural laws can be properly understood only if one acknowledges the existence of universals. The gist of this argument is that an opponent of universals can at best represent natural laws as consisting merely in universal constant conjunctions amongst particulars, which reduces those laws to nothing more than cosmic coincidences or accidents. The remedy, it is then proposed, is to regard

---

[14] See further my 'Locke, Martin and Substance'.

natural laws as consisting in relations between *universals* rather than in constant conjunctions amongst particulars and, more specifically, as involving a 'second-order' relation of 'necessitation' between universals.[15] Thus, in the simplest sort of case, it is proposed that the form of a law is something like '*F*-ness necessitates *G*-ness'. It is then further claimed that although '*F*-ness necessitates *G*-ness' entails the corresponding constant conjunction amongst particulars, namely, 'For any (individual) x, if x is F, then x is G', the reverse entailment does not hold, so that the law is stronger than the constant conjunction and *explains* it, thus differentiating such a conjunction which is backed by a law from one which is a mere cosmic coincidence—a differentiation which the opponent of universals is unable to make or explain.

All of this is familiar territory, as is the objection raised in some quarters that the supposed entailment of 'For any x, if x is F, then x is G' by '*F*-ness necessitates *G*-ness' is unexplained and mysterious.[16] To this I would add the objection that laws do *not*, in fact, entail constant conjunctions amongst particulars in any case, because laws—apart, perhaps, from certain fundamental physical laws—admit of exceptions which arise from the possibility of interfering factors in the course of nature, an example being the possible deviation of planets from their elliptical orbits as specified by Kepler's laws of planetary motion.[17] Laws, in my view, determine *tendencies* amongst the particulars to which they apply, not their actual behaviour, which is a resultant of many complex interactions implicating a multiplicity of laws.

However, this is not the main source of my dissatisfaction with the view that laws have the form, in the simplest sort of case, '*F*-ness necessitates *G*-ness'. For, although I certainly agree that the key to understanding the ontological status of laws is to recognize them as involving universals rather than particulars, I consider that we can only understand laws properly if we recognize as ontologically fundamental the distinction between substantial and non-substantial universals. Looking back at the diagram introduced earlier (Fig. 2.1), we can see directly from this in what way a law involves universals. A law simply consists—in the simplest sort of case—in some substantial universal or kind being characterized by some non-substantial universal or property, or in two or more kinds being characterized by a relational universal. Our very statements of law in everyday and scientific language tell us this, if only we are prepared to take them at their literal face value. We say, for example, 'Rubber stretches', 'Gold is ductile', 'Water dissolves common salt', 'Planets move in elliptical orbits', 'Electrons carry unit negative charge' and 'Protons and electrons attract each other'. These are all statements of

---

[15] See, especially, D. M. Armstrong, *What Is a Law of Nature?* (Cambridge: Cambridge University Press, 1983).

[16] For this complaint, see Bas van Fraassen, *Laws and Symmetry* (Oxford: Clarendon Press, 1989), ch. 5.

[17] See again my *Kinds of Being*, ch. 8, and also my 'What *is* the "Problem of Induction"?', *Philosophy* 62 (1987), pp. 325–40.

natural law and each can be understood, in terms of the ontological system represented in our diagram, as saying that one or more substantial kinds is or are characterized by some property or relation. There is no need, then, to think of the universals involved in a law as being linked by some mysterious 'second-order' relation of 'necessitation', for they are in fact linked quite simply by the familiar characterizing tie which links substantial entities quite generally, whether universal or particular, to non-substantial entities. The basic form of a law is not '*F*-ness necessitates *G*-ness', but '*K*s are *F*', or '*K*s are *R*-related to *J*s', where '*K*' and '*J*' denote substantial universals, '*F*' denotes a property and '*R*' denotes a relation—that is, where '*F*' and '*R*' denote non-substantial universals.

Lest it be complained that the 'characterizing tie' invoked here is itself mysterious or problematic, I hasten to point out that this is not an objection which can be levelled against me by those philosophers who accept universals but merely dispute my differentiation of them into substantial and non-substantial universals—that is to say, the kind of philosophers who support the view that laws have the form '*F*-ness necessitates *G*-ness'. For these philosophers too are committed to the existence of some sort of characterizing tie, either directly between individual substances and universals or between individual substances and property-instances. Nor do I, in any case, accept that the notion of the characterizing tie, properly understood, is inherently mysterious or problematic. The key to understanding it is not to regard this 'tie' as a genuine *relation*, for once we do that we are doomed to set out upon the sort of regress made famous by Bradley. Rather, we need to appreciate—adapting Frege's terminology to a slightly different purpose—the essential 'unsaturatedness' of non-substantial entities as being what makes such entities necessarily 'tied' to substantial ones.[18] This is why, earlier, I described characterization, along with its relatives instantiation and exemplification, as 'relationships' rather than as 'relations'. What we are concerned with here are certain species of metaphysical dependency, not *relational universals* in the sense in which these are members of one of our fundamental ontological categories of *being* or *entity*.

2.5 DISPOSITIONAL VERSUS OCCURRENT PREDICATION

With this account of the ontological status of natural laws in place, the four-category ontology now delivers a further important insight which confirms its fruitfulness as a metaphysical hypothesis. This is that it enables us to understand in a fresh way the much-bedevilled distinction between the dispositional and the occurrent, or 'categorical', features of objects—although I shall avoid the expression 'categorical'

---

[18] See further my 'Locke, Martin and Substance'. However, as we shall see in later chapters of this book, we need to go beyond the metaphor of 'saturation' for an account that is fully perspicuous.

as potentially misleading, especially in the context of the present book. Some philosophers have regarded this as a distinction between types of *property*, but that seems to me to be a serious mistake. Others have held that it is a distinction between different 'aspects' which every property possesses, which is a more sustainable view but still, I think, mistaken.[19] I prefer to start with a distinction between dispositional and occurrent *predication*, a distinction which is reflected in language but which at bottom, I believe, rests on an ontological distinction—one that is immediately apparent from the diagram presented earlier.[20] It will be observed from that diagram that there are two fundamentally different ways in which an individual substance can be related ontologically to—that is, in which it can *exemplify*—a non-substantial universal. One way is for the individual substance to be characterized by a property-instance which instantiates the non-substantial universal in question. The other way is for the individual substance to instantiate a substantial universal or kind which is characterized by the non-substantial universal in question. These two ways correspond, I contend, to occurrent and dispositional predication respectively, in the sense that these forms of predication can be understood as having as their respective truthmakers the two types of circumstance just described.

Consider, for example, a dispositional predication such as 'This object is soluble in water' or, as we might also put it, 'This object *dissolves* in water', and contrast this statement with the corresponding occurrent predication, 'This object *is dissolving* in water'. Many fruitless attempts have been made to explain or analyse such a dispositional predication, sometimes by reducing it to some sort of conditional statement involving only occurrent predication, along the lines of 'If this object were to be immersed in water, then it would be dissolving in water'. These attempts have all foundered on various obstacles and, although some philosophers still desperately try to remedy their defects, it should be clear by now that the strategy is doomed to failure.[21] However, the four-category ontology offers a simple explanation of the distinction that underlies the difference between these two types of predication, the dispositional and the occurrent. To say of an object that it is disposed to dissolve in water, according to my proposal, is effectively to say that the object instantiates some kind (that is, some substantial universal), K, such that water dissolves K—where, of course, 'Water dissolves K' expresses a natural law, as explained earlier. An example of such a law would be the previously cited law, 'Water dissolves common salt'. And, as was also explained

---

[19] For this view, see C. B. Martin's contributions to D. M. Armstrong, C. B. Martin and U. T. Place, *Dispositions: A Debate*, ed. Tim Crane (London: Routledge, 1996). See also C. B. Martin and John Heil, 'The Ontological Turn', *Midwest Studies in Philosophy* XXIII (1999), pp. 34–60.

[20] For an early version of the theory of dispositions about to be described, see my 'Laws, Dispositions and Sortal Logic', *American Philosophical Quarterly* 19 (1982), pp. 41–50, and my *Kinds of Being*, ch. 9.

[21] See, especially, C. B. Martin, 'Dispositions and Conditionals', *Philosophical Quarterly* 44 (1994), pp. 1–8. This paper has already generated a considerable literature but, to my mind, no satisfactory response in defence of a conditional analysis of disposition statements.

earlier, laws only determine the *tendencies* of individual substances, not their *actual* behaviour, which has a multiplicity of determinants, including the actual behaviours of many other individual substances. The other half of the proposal is that the corresponding *occurrent* predication, in which it is said that the object is (actually) dissolving in water, effectively says of the object that it possesses a (relational) *property-instance* of the non-substantial universal *being dissolved by water*. Similarly, to say of a particular piece of rubber that it *is stretching* is to say that it possesses an instance of the very same non-substantial universal which, in virtue of the law 'Rubber stretches', characterizes the substantial universal or kind *rubber*: whereas to say that the piece of rubber *stretches*—in other words, that it is elastic—is to say that it instantiates *some* kind which is characterized by the non-substantial universal in question.

I shall have much more to say about these matters in later chapters of this book. But what should be clear, at any rate, is that if the ontology is as our diagram depicts it, then these two modes of predication ought to be available, just as we apparently find them in fact to be. So the theory predicts and explains the existing linguistic phenomena, which is to its credit—whereas no rival theory has, I believe, come anywhere near to achieving the same result. Other theories have so far left the distinction between the dispositional and the occurrent altogether mysterious and obscure.

2.6 ADVANTAGES OF THE FOUR-CATEGORY ONTOLOGY

Let me now sum up some of the advantages of the four-category ontology over its various more parsimonious rivals. Its advantages over a one-category ontology of property- and relation-instances (a pure trope ontology) are (1) that it can provide, as this rival cannot, an adequate account of the existence and identity conditions of property- and relation-instances, (2) that it has a superior account of the ontological status of natural laws, and (3) that it offers a principled

Table 2.1. Advantages of the four-category ontology

	Trope individuation	Analysis of laws	Analysis of dispositions	Property perception
Tropes alone	No	No	No	Yes
Tropes plus objects	Yes	No	No	Yes
Objects plus universals	Not applicable	Yes	No	No
Four-category ontology	Yes	Yes	Yes	Yes

understanding of the distinction between the dispositional and the occurrent. Its advantages over a two-category ontology of individual substances and property- and relation-instances are the second and third of the aforementioned advantages. And its advantages over a two-category ontology of individual substances and non-substantial universals is the third of the aforementioned advantages, together with (4) the advantage of providing a superior account of our empirical acquaintance with the properties and relations of individual substances, which avoids mysterious talk of the 'multiple location' of universals. These advantages may be summed up in Table 2.1.

3

Some Formal Ontological Relations

3.1 DEPENDENCE RELATIONS AND THEIR FOUNDATION

Any system of ontology must embody an account of the ontological categories—that is, an account of what an ontological category is, what ontological categories there are, and which ontological categories are fundamental, as well as an account of which ontological categories are actually occupied and which are not. In giving such an account, it will be necessary for the ontologist to specify certain *formal ontological relations* between entities belonging to the same or different ontological categories. Different ontologists will, of course, have different views as to which formal ontological relations need to be invoked for these purposes. My own view is that at least the following different such relations need to be invoked: *identity, instantiation, characterization, exemplification, constitution, composition*, and *dependence*. Of these, dependence is not so much a single relation as a family of relations, including, for example, *rigid* existential dependence, *non-rigid* existential dependence and also (what I call) *identity dependence*. Moreover, as I shall shortly try to explain, it would seem that all dependence relations are, in a certain sense, *founded* upon other formal ontological relations—relations which are, for this reason, ontologically more basic than the dependence relations themselves.

To illustrate some of the differences between species of dependence, here are some examples. Rigid existential dependence is exemplified by the dependence of a boundary or a hole upon its 'host' (that is, the thing that has the boundary or in which the hole exists) or, again, by the dependence of a heap of stones upon the individual stones that it contains. Non-rigid existential dependence is exemplified by the dependence—at least according to an 'Aristotelian' or immanent realist view of universals—of a universal upon its particular instances. And identity dependence is exemplified by the dependence of a unit set upon its sole member or—according to some substance ontologists, at least—by the dependence of a property-instance or 'mode' upon the individual object to which it 'belongs', or in which it 'inheres'.

Rigid existential dependence may be formally defined as follows:

x depends rigidly on $y =_{df}$ necessarily, x exists only if y exists

I take it that identity dependence entails rigid existential dependence: that if one entity depends for its identity upon another, then the former could not have existed without the latter. For example, it is because a hole depends for its identity upon its host—in the sense that *which* hole it is is at least partly determined by *which* object its host is—that the hole could not have existed without the host. But the reverse entailment does not hold. Thus, although no entity could have existed without the unit set of which it is the sole member, the identity of an entity is not determined by the identity of its unit set, but quite the reverse. Identity dependence is asymmetrical, whereas rigid existential dependence is not.[1] As for *non-rigid* existential dependence, I shall say more about that in a moment.

It will be noticed that in giving these examples, I have invoked certain formal ontological relations besides dependence relations as such. Thus, in giving the example of a universal and its particular instances, I implicitly invoked the relation of *instantiation* in which those instances stand to the universal. And in giving the example of an individual object and its modes, I invoked the relation of *inherence*—or, as I prefer to call it, *characterization*—in which those modes stand to the object. Indeed, in giving the example of a unit set and its sole member, I also implicitly invoked the relation of *set-membership* in which a member of a set stands to that set. I did not include the relation of set-membership in my original list of indispensable formal ontological relations only because I have some doubts as to its indispensability. This in turn is because I have some doubts concerning the degree to which we should regard sets with full ontological seriousness.[2] It may be, perhaps, that talk of sets can be reconstrued as talk of pluralities that are 'nothing over and above' the several entities comprising them and that talk of such an entity 'being a member of' a set can correspondingly be reconstrued as talk of that entity 'being one of' the entities in question. Is 'being one of' a plurality of entities a *relation* in which something may stand to the entities in question? Not, it might seem, in any fundamental or irreducible sense. After all, something is one of the intrajovine planets just in case it is identical with either Mercury, Venus, Earth, or Mars. *Identity*, however, figured on my list of indispensable formal ontological relations and in this case, at least, we have eliminated mention of the putative 'being one of' relation in favour of talk of identity.

I have included this little discussion of the ontological status of sets at this point just to illustrate one sort of way in which an ontologist might reasonably seek to reduce the number of formal ontological relations to a select few. It doesn't matter to me very much for present purposes whether the example illustrates a *successful* attempt to do this sort of thing. But I take it that an ontologist should not

---

[1] For a fuller discussion of rigid existential dependence and identity dependence, including a formal definition of the latter, see my *The Possibility of Metaphysics: Substance, Identity, and Time* (Oxford: Clarendon Press, 1998), ch. 6.
[2] See further my *A Survey of Metaphysics* (Oxford: Oxford University Press, 2002), p. 377. But, for a less sceptical view, see also my *The Possibility of Metaphysics*, pp. 220 ff.

multiply formal ontological relations beyond necessity and that an ontological system which invokes fewer such relations is, other things being equal, preferable to one which invokes more.

Let us now look more closely at the example of *non-rigid* existential dependence offered earlier, that of the non-rigid existential dependence of an (Aristotelian or immanent) universal upon its particular instances, such as the non-rigid existential dependence of the universal *doghood*, or *being a dog*, upon individual dogs. The thought here is that, necessarily, this universal exists only if *some* individual dogs exist, even though the universal does not depend rigidly for its existence upon any individual dog 'in particular' (as we say). Supposing, for instance, that Fido and Rover are the only existing dogs, we do not want to imply that doghood would not have existed if neither Fido nor Rover had existed, but only that it would not have existed if no individual dog whatever had existed. None the less, it seems that there is a perfectly good sense in which doghood does *actually* depend for its existence upon Fido and Rover, because it depends for its existence on there being *some* individual dogs and it turns out that Fido and Rover are (as we suppose) all the individual dogs that there actually are. And it is this sort of existential dependence that may aptly be called 'non-rigid'.

It would seem that non-rigid existential dependence may, accordingly, be formally defined as follows:

x depends non-rigidly on the ys $=_{df}$ for some F, the ys are the Fs and, necessarily, x exists only if there is something z such that z is an F

To apply this definition to our example, let x be doghood, let the ys be the dogs Fido and Rover, and let F be *instance of doghood*: then doghood depends non-rigidly on the dogs Fido and Rover, according to our definition, because Fido and Rover are (as we suppose) the instances of doghood and, necessarily, doghood exists only if there is something that is an instance of doghood.

Obviously, as a special case of non-rigid dependence, we have the non-rigid dependence of x on a single entity, y, such that for some F, y is the F. Note, moreover, that 'x depends rigidly on y' entails 'x depends non-rigidly on y', according to our definitions. This is because if x depends rigidly on y, then there is some F—namely, *entity that is identical with y*—such that y is the F and, necessarily, x exists only if there is something z such that z is the F. That our definitions should have this implication would seem to be to their credit, since rigid dependence is, intuitively, a stronger relation than non-rigid dependence. However, it must be acknowledged that in order to avoid triviality in our definition of non-rigid dependence, some suitable restriction must be imposed on the admissible values of F, because without such a restriction the definition implies that, for any ys whatever, x depends non-rigidly on the ys. For example, it implies that x depends non-rigidly on any arbitrarily chosen entities a and b, if we are allowed to let F be something like *entity that is identical with a or b, or with something other than a or b only if neither a nor b exists*. I leave to another occasion the task of specifying the required restriction.

None of this seems particularly problematic. What may, however, need some clarification is the relationship between non-rigid existential dependence and instantiation implicit in our example of doghood and the dogs Fido and Rover. Fido and Rover each stand in the instantiation relation to the universal doghood. At the same time, doghood stands, it seems, in the non-rigid existential dependence relation to Fido and Rover. Does this mean, then, that instantiation is somehow 'nothing over and above' non-rigid existential dependence? In which case, do we really need to include instantiation as such on our list of indispensable formal ontological relations? A similar question arises with regard to characterization (or 'inherence') and *rigid* existential dependence. A substance ontologist may hold that an individual round ball is characterized (at any given moment of time at which it is round) by a particular roundness, which could not have existed save as the roundness of that very ball and which, hence, stands in the relation of rigid existential dependence to the ball. But then we seem to be talking of *two* relations between the roundness and the ball: characterization (or 'inherence') and rigid existential dependence. Do we really need to invoke *two* relations, however, or will it suffice to invoke just *one*—in which case, *which* one?

In response to such queries, I am inclined to reply as follows. As a substance ontologist and advocate of immanent realism concerning universals, I do not believe that the relations of characterization and instantiation are dispensable, but I am also committed to certain ontological dependency claims associated with these positions and consequently, I believe, to the indispensability of various dependence relations. However, I regard the rigid existential dependence relation between an individual object and one of its modes as being *constituted* by the characterization relation in which the mode stands to the object. Likewise, I regard the non-rigid existential dependence relation between a universal and its particular instances as being *constituted* by the instantiation relations in which the instances stand to the universal. And *constitution*, of course, is itself one of the formal ontological relations that I included on my original list of indispensables. More generally, on this way of viewing the matter, all dependence relations between entities will be constituted by certain other formal ontological relations in which those entities stand to one another, so that these other formal ontological relations will always be, in a certain sense, more fundamental than the various dependence relations that they constitute in various different cases.

If I am right about this, there are no 'brute' or 'unconstituted' dependence relations between entities. This is one reason why I placed 'dependence' at the end of my list of formal ontological relations, for although I think that dependence relations are perfectly genuine, I do not think that they are basic. In saying that they are 'genuine', I mean that we cannot regard talk of them as being on a par with talk of the 'being one of' relation, on the assumption that talk of the latter kind is eliminable in favour of talk in terms of identity in the way suggested earlier.

3.2 ONTOLOGICAL CATEGORIES AND THEIR ORGANIZATION

I began this chapter by saying that we need to invoke various formal ontological relations in developing any system of ontology, but also that there is much more that we need to do for this purpose. Most importantly, we need to invoke our various formal ontological relations in the context of an account of the *ontological categories*. So far I have been invoking exemplars of various ontological categories in an unsystematic way for purposes of illustration, but now we need to approach the issue of categorization more systematically. There is, of course, no general consensus of opinion as to what ontological categories we should recognize and how exactly we should think of them as being organized, although it is widely accepted that some sort of hierarchical principle of organization is correct. For present purposes I am going to follow previous chapters and assume without further argument that the top level of an adequate system of ontological categories will look as follows.[3] The topmost category is the category of *entities* or *beings*, to which anything whatever uncontentiously belongs. At the next immediate level, there is an exhaustive and exclusive division between *universals* and *particulars*. Universals then divide into two sub-categories: *properties and relations* on the one hand and *kinds* on the other. The particular instances of kinds are individual *objects*, whereas the particular instances of properties and relations are, respectively, monadic and relational *modes*.[4]

There are various sub-categories of objects, by far the most important being that of individual *substances* (which is why, elsewhere, I often use the terms 'particular object' and 'individual substance' interchangeably, when this will not cause confusion). Amongst the other sub-categories of objects one finds such things as holes and heaps. According to this scheme, such entities as events and processes (sometimes called 'occurrents', in contrast with 'continuants'), either find their proper place amongst the modes or else should somehow be understood as being constituted by 'complexes' which include objects and universals amongst their constituents. So, for example, a particular collision might either be identified simply with a particular *colliding*—a relational mode—or else be thought of as consisting in the exemplification of the corresponding relational universal at a certain time by two or more material objects.[5]

Thus, the topmost levels of the categorial hierarchy will look like the diagram in Fig. 3.1.

---

[3] See also my *The Possibility of Metaphysics*, pp. 179 ff and my *A Survey of Metaphysics*, pp. 13 ff.

[4] For more on the distinction between properties and modes and why it needs to be acknowledged, see Chapter 6 below, which is based on my 'Properties, Modes and Universals', *The Modern Schoolman* 74 (2002), pp. 137–50.

[5] I say more about the nature of events in my *A Survey of Metaphysics*, ch. 13. See also Chapter 5, section 5.7, below.

Some Formal Ontological Relations 39

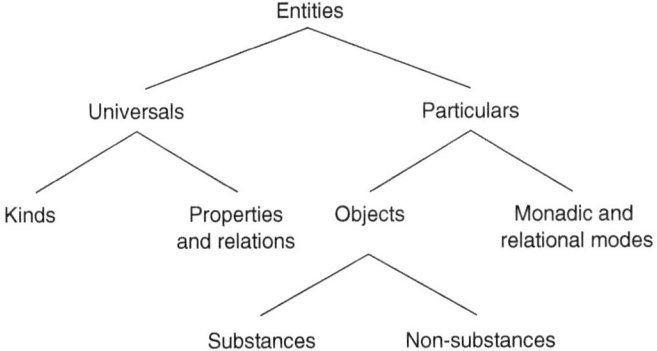

Figure 3.1. Topmost levels of the categorial hierarchy

As readers should by now be well aware, I call this system the *four-category ontology*. Not, of course, because it contains only four ontological categories, for it plainly does not. Rather, I call it this because it contains just four *fundamental* ontological categories, which appear at the third level of categorization, these categories being those of *kinds*, *properties* (and relations), *objects*, and *modes*. The higher level categories of *universal*, *particular*, and *entity* are not more fundamental than those of the third level because they are mere abstractions and do no serious ontological work on their own account. This is perhaps obvious in the case of *entity*, since nothing whatever is excluded from this category. As for the distinction between *universal* and *particular*, I simply define it in terms of the instantiation relation. A particular is that which has (or, in a stronger version, that which *can* have) no instances, whereas a universal is that which has (or, in a weaker version, that which *can* have) instances.[6] But, although this is an important distinction, it serves to explain nothing in ontology that is not fully explicable in terms of the defining features of the four categories at the third level.

The defining features of the four categories at the third level are as follows. Kinds are those universals that have objects as their instances. Properties and relations are those universals that have modes as their instances. Objects are characterized by modes. Kinds are characterized by properties and relations. In specifying these defining features I have, of course, invoked another formal ontological relation in addition to instantiation, namely, characterization (or 'inherence'). As we saw in Chapter 2 above, we can depict these features in the form of a diagram (see Fig. 3.2), which may be called, suitably enough, the Categorial or Ontological *Square*.

As was remarked in Chapter 2, the Ontological Square leaves room for two different ways in which objects may be related to properties and relations: either

---

[6] See further my *A Survey of Metaphysics*, pp. 350 ff.

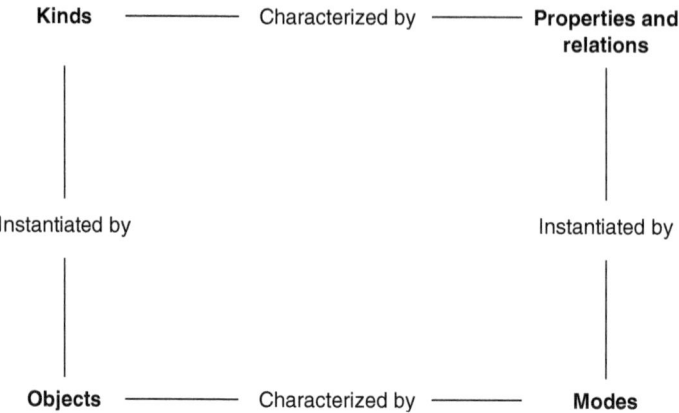

Figure 3.2. The Ontological Square (II)

(1) by their instantiating kinds that are characterized by those properties and relations or (2) by their being characterized by modes which instantiate those properties and relations. I call both of these two different relations (species of) *exemplification* and contend that the difference is reflected in natural language in terms of the distinction between dispositional and occurrent predication. To say that a rubber ball is 'occurrently' round is, thus, true just in case the ball is characterized at the time of utterance by a mode that is an instance of the property of roundness or being round. In contrast, to say that the ball is 'dispositionally' round—as one might, even at a time when, owing to its being compressed, it is not occurrently round—is true just in case the ball itself instantiates a kind that is characterized by the property of roundness, that is, in more idiomatic language, just in case the ball is a kind of thing that is round by nature. Exemplification, then, is a relation that comes in two different varieties or species, but in each case it is *constituted* by a certain combination of two other relations—instantiation and characterization—so that these latter relations should in all cases be deemed to be more fundamental than exemplification. (If one wanted to include exemplification in the diagram representing the ontological square, one would have to represent it by a diagonal reaching from the bottom left-hand corner to the top right-hand corner and see this as the resultant of either of two different routes that could be taken around the sides of the square: see Fig. 1.2 of Chapter 1 above.)

3.3 THE ONTOLOGICAL STATUS OF THE CATEGORIES

I have just been talking extensively about ontological categories and formal ontological relations. But amongst the categories that I have identified, at the third level, are the category of *kinds* and the category of *properties and relations*.

This fact, however, prompts the following question. Do the ontological categories themselves belong to, say, the category of kinds and do formal ontological relations belong to the category of properties and relations? In other words, are ontological categories and formal ontological relations themselves *universals*—as opposed, of course, to *particulars*? They could only be so, of course, if they were *entities*. But are they? And if we say that they are not, how can we contend that our ontology is a realist one? Now, it may seem to be blatantly self-contradictory to say that formal ontological relations are *not* relations, but I shall postpone investigation of this matter until I have dealt with the seemingly less clear-cut case of the ontological categories, since what we say in the latter case may be able to help us in the former.

The first thing I want to say with regard to the ontological categories is this— that they are certainly *not* universals. Indeed, it seems that they plainly couldn't be. Why not? Because the defining feature of universals, I have suggested, is that they have (or at least can have) particulars as their instances. For example, the property of being round is a universal, because it may have some object's particular roundness as an instance. Again, the kind *dog* is a universal, because it may have some particular dog as an instance. But suppose, now, that we said that all ontological categories are universals. Then, more specifically, we should have to say that the first of the four third-level categories, that of *kinds*, is a universal. But if so, it should have, or be capable of having, particular instances. What could those instances be? Plainly, they could only be the various *kinds*, such as the kind *dog*. But the kind *dog* and all other kinds are necessarily *universals* and so *not* particulars. So the category of kinds cannot have particulars as instances and hence cannot qualify as a universal. (I shall deal in a moment with the question of whether the category of kinds might be a so-called 'higher-order' universal.)

Must the category of kinds then qualify as a *particular*? Certainly it must, if it is an entity but not a universal and the distinction between universals and particulars is exhaustive and exclusive. However, the defining feature of particulars, I have suggested, is that they cannot have instances. So what is the relation between the category of kinds and the various kinds that 'belong' to that category, if it is not the relation of instantiation? Is it perhaps the relation of set-membership, so that the category is a set (an abstract *object*) and the kinds are its members? (Let us waive here my earlier doubts as to whether we should treat sets with full ontological seriousness.) If we were to adopt this position with regard to the category of kinds, namely, to regard it as a *set*, then it would seem that consistency would require us to treat all ontological categories in the same manner. On this view, then, all of the ontological categories are sets, including the category of *entities*, and everything that there is belongs, in the set-membership sense of 'belongs', to one or more of these sets. But everything, we have said, belongs to the category of *entities*, so that if this category is a set, it is one which has itself as a member. Now, however, we are apparently in deep water, from which we would do well to extricate ourselves as quickly as possible, because it is, with good reason, standardly assumed that no set can be a member of itself. (The same problem arises with the category of *objects*, because if this is a set and thus an abstract *object*, then it must belong to itself.)

One possibility that has not so far been explored is to regard the category of kinds, and perhaps all ontological categories, as being *higher-order* universals—where a first-order universal is understood as before as being something that has (or can have) particulars (objects or modes) as instances, a second-order universal is something that has (or can have) first-order universals as instances, and so on. A putative example of a second-order universal would be the (putative) property of being a colour-property, whose (putative) instances are such first-order properties as the property of being red and the property of being green. (On reflection it might seem more appropriate to think of the putative higher-order universal of which these first-order colour-properties are supposedly instances as being a higher-order *kind* rather than a higher-order *property*. But the difficulty here is that, since the instances of kinds are supposed to be 'objects', we would then seem to be treating first-order properties as being *second-order objects* —and it is not at all clear what sense can be made of this.)

On this view, then, the category of kinds would be, perhaps, the second-order property of *being a kind*, whose instances are all the various *kinds*, such as the kind *dog*. But what now about, say, the category of *objects* (objects being understood, as hitherto, as particular instances of kinds)? Would this likewise be a second-order property? Presumably not: presumably it would have to be the *first-order* property of *being an object*, whose instances are all the various *objects* (although there is an immediate difficulty with this suggestion to which I shall return shortly). So it will turn out that, in one sense, not all categories belong to the same category: the category of kinds will belong to the category of second-order universals while the category of objects will belong to the category of first-order universals—even if they do both qualify as universals and so in *that* sense 'belong to the same category'. But what now about the category of universals itself? If that is a universal, of what order is it? It is hard to see how it could be of any order, because a universal of any order whatever must belong to the category of universals—and so if the category of universals is a universal of some order, it seems that it must, absurdly, be of a higher order than itself. (To all of this I would add that I am, in any case, sceptical about the need to acknowledge the existence of any 'higher-order' universals whatever, which is why I have appended none to the Ontological Square.)

There is a further difficulty with the preceding suggestion that the category of objects is a (first-order) *property*, the property of being an object, which has all the various objects as its particular instances—and this is that, according to the four-category ontology, the particular instances of (first-order) properties are *modes*, not objects. It would seem, then, that if we are to say that the category of objects is a (first-order) universal, we shall have to say that it is a *kind*—as it were, the 'highest' kind to which all objects belong. (Kinds have their own hierarchy, which is quite distinct from the hierarchy of the ontological categories: for instance, the kind *dog* is a sub-kind of the kind *mammal*, which is in turn a sub-kind of the kind *animal*.) But, plausibly, there is no reason to think that there is a 'highest kind' to which all objects whatever belong, because objects receive their identity conditions from the kinds to which they belong and yet objects of different kinds may have quite different identity conditions—for example, the identity conditions of animals are different from those of artefacts and different again from those of atoms.[7] Hence, a putative 'highest kind'—the kind *object*—could not supply identity conditions for all of its instances and so wouldn't genuinely qualify as a 'kind'.

As far as I can see, the only acceptable thing to say, in view of all the preceding difficulties, is that the ontological categories are not themselves entities and are thus not to be included in an exhaustive inventory of what there is. There are, quite literally, no such things as ontological categories. (That indeed is why, in the tree of categories depicted earlier, it was entirely appropriate to label the nodes by *plural* terms, such as 'entities' and 'kinds', for these terms do all genuinely refer—plurally—to entities, the first obviously so. In contrast, it would have been misleading to label the nodes by *singular* terms, such as 'the category of entities' and 'the category of kinds', given that these do not refer to anything at all.)

Does this mean that no ontological system can have a realist foundation? Does it mean that ontological categorization is all just a matter of how we choose to classify and describe things—of how we choose to 'carve up' reality, to use the rebarbative metaphor so often favoured by antirealists? Not at all. The difference between, say, an object and a property, or between an object and a mode, is as fundamental, objective and real as any difference could possibly be. Such differences do not rest purely on our ways of describing or conceiving of things, despite the spurious arguments to the contrary advanced by some philosophers, such as Frank Ramsey.[8] Ramsey, it will be recalled, suggested that the distinction between universal and particular is purely linguistic in origin and has no ontological significance, on the grounds that 'Socrates is wise' can be paraphrased as 'Wisdom is a characteristic of Socrates', in which the subject and predicate of the original sentence have their roles reversed. But this sort of argument would only have any force against someone who mistakenly tried to define the distinction between universal and particular in linguistic terms, by saying something to the effect that universals are what are denoted by predicative expressions and particulars are what are denoted by subject terms.

One can be a realist concerning the distinctions that are captured by a system of ontological categories without having to maintain that the categories themselves are elements of being. An object is different from a property or a mode in virtue of the intrinsic natures of these entities, quite independently of us and our ways of

---

[7] See further my *Kinds of Being: A Study of Individuation, Identity and the Logic of Sortal Terms* (Oxford: Blackwell, 1989), ch. 2.
[8] See F. P. Ramsey, 'Universals', in his *The Foundations of Mathematics and Other Logical Essays* (London: Kegan Paul, 1931). I discuss Ramsey's arguments much more fully in Chapter 7 below.

describing or thinking of things. We place things in different ontological categories correctly if we distinguish them rightly in respect of these intrinsic and objective differences between them. Categorizing things correctly differs, however, from other ways of classifying things (for example, scientific taxonomies of biological species or chemical elements), insofar as it rests upon considerations that are at once highly abstract or general in character and amenable to purely *a priori* evaluation. More specifically, we categorize correctly when we categorize by correctly apprehending the *existence and identity conditions* of the things concerned. For example, what is distinctive of a *mode* is that it depends for its existence and identity upon the object that it characterizes and that it instantiates a property—a universal—that is non-rigidly existentially dependent upon the modes that are its particular instances. A mode, thus understood, is something intrinsically and objectively different from anything belonging to, say, the category of *objects*. Thus, there could be no distinction more fundamental or mind-independent than that between, for example, a round ball and that same ball's roundness. This would still be true even if we were to decide, for independent reasons, that there are in fact no material objects and so no round balls. For, although I believe that it is necessarily the case that all four of the fundamental ontological categories are actually occupied, there is plenty of scope to debate whether or not various sub-categories of those basic ones are filled in actuality.

3.4 ARE FORMAL ONTOLOGICAL RELATIONS ELEMENTS OF BEING?

Now we are in a position to tackle the more difficult question of whether formal ontological relations are indeed *relations*, in the sense of belonging to the sub-category of universals that I have identified as the category of properties and relations. To get a proper grip on the question, we need to think again about the defining features of the entities that belong to this category. Properties and relations, I have said, are universals and as such have (or at least can have) particular instances, these being monadic or relational modes. The modes that are instances of properties and relations are entities that characterize (or 'inhere in') objects and they depend for their existence and identity upon those objects. So let us then consider a formal ontological relation, such as instantiation or characterization, and ask whether or not we can coherently suppose it to qualify as a *relation*, given what has just been said about such an entity.

Consider, for instance, the characterization relation in which a mode stands to the object that it characterizes (or in which it 'inheres'), as in the example of the round ball and its roundness. If characterization is really a relation—that is, if it is a relational universal—then it must have particular instances, these being relational modes. So, it seems, we must say that, inasmuch as the ball's roundness characterizes the ball at a given time, the ball and the roundness are related by a relational mode of characterization, that is, by a particular instance of the characterization relation. Whereas we might have thought, then, that we were confronted with just *two* particulars in this case—the ball and its roundness—it turns out that we are confronted with *three*: the ball, its roundness, and the relational characterization mode 'between' them. Remember, however, that monadic and relational modes are supposed to characterize *objects*—either single objects, in the case of monadic modes, or pluralities of objects in the case of relational modes. But the relational characterization mode that we have just postulated cannot be understood to play this sort of role, because it is supposed to relate not two or more *objects*, but a monadic mode (the ball's roundness) and a single object (the ball). Hence, it cannot properly qualify as a mode. If it exists at all, it belongs to some other, quite distinct, ontological category. The four-category ontology certainly doesn't allow for the existence of any such entity, however, so either no such entity exists or else the four-category ontology is defective.

Before we attempt to pass a verdict on this issue, let us consider another example, that of the formal ontological relation of *instantiation*, and ask again whether this can really be thought of as being a relational universal. It should be evident that we are going to find ourselves in difficulties similar to those that arose in the case of characterization, if we seek to defend a positive answer to this question within the framework of the four-category ontology. Fido, a particular dog, instantiates the kind *dog*. How are we to think of the 'relation' of instantiation in this context? According to the four-category ontology, what relational universals relate are *kinds*, which they do in virtue of *characterizing* those kinds. For example, the relation of *being heavier than* characterizes the kinds *elephant* and *mouse*, taken in that order. As we ordinarily put it: elephants are heavier than mice. On this conception, relational universals cannot relate a particular, such as Fido, and a universal, such as the kind *dog*. Of course, there is also a derivative sense in which the relation of being heavier than relates two particulars, such as, perhaps, the two dogs Rover (an Irish wolf hound) and Fifi (a miniature poodle), inasmuch as a particular instance of the relation in question characterizes those two dogs, taken in that order. (Although, as we shall shortly see, our choice of example may be unfortunate if the relation of being heavier than is to be understood as being an 'internal' one.) But, again, this is not a case which helps in any way to make sense of the suggestion that instantiation is a relational universal that relates a particular to a universal.

So, it seems, the four-category ontology cannot accommodate the suggestion that formal ontological relations, such as characterization and instantiation, are genuine *relations*—that is, that they are elements of being (entities) belonging to the category of universals.[9] Does this imply a fatal deficiency in the four-category

---

[9] A somewhat similar view is taken by David Armstrong: see his *A World of States of Affairs* (Cambridge: Cambridge University Press, 1997), p. 118.

ontology? I believe not, any more than I believe it to be a deficiency in that ontology that it can make no sense of the suggestion that the categories themselves are elements of being. But how can we even talk about 'the ontological categories' and 'the formal ontological relations' without supposing that they are *something*—and hence that they are entities that should figure somewhere in the hierarchy of ontological categories? As a first defensive move, we can, of course, point out that not every meaningful predicate, whether monadic or relational, need or indeed can be supposed to denote an existing property or relation—so that it is not obligatory to suppose that predicates such as 'is a universal' or 'instantiates' denote, respectively, a monadic and a relational universal.[10] However, if we are to adhere to the four-category ontology, we need to treat this kind of move with caution, as we obviously don't wish to apply it with respect to those predicates that we *do* regard as denoting universals, such as, perhaps, 'is square' and 'loves'. After all, the four-category ontology is explicitly opposed to any form of nominalism.

One strategy that is sometimes favoured with regard to such seeming relations as instantiation and characterization is to classify them as 'internal' relations, understanding an internal relation to be one which supervenes upon the natures (or, more generally, upon the intrinsic properties) of its relata.[11] The relation of being taller than is said to be internal in this sense, because whether or not a is taller than b is determined entirely by the heights of a and b—whereas, for example, whether or not a is five metres away from b is not, apparently, similarly determined entirely by the 'intrinsic' properties of a and b. On this account, it is sometimes said that a's being taller than b involves 'no addition of being' to a's being the height it is and b's being the height it is. In a similar fashion, it may be said that whether or not Fido instantiates doghood is entirely determined by the natures of Fido and doghood and that whether or not a certain roundness characterizes a certain ball is entirely determined by the natures of the roundness and the ball.

On this account, then, it may likewise be said that the 'relations' of instantiation and characterization involve 'no addition of being' and that this is why they do not need to be regarded as relational universals.[12] In a way, these 'relations' are thus treated as being nothing in themselves, which accords well with my earlier refusal to acknowledge them as genuine 'entities'. On the other hand, there seems to be something very different about the case of instantiation and characterization from the case of, say, the relation of being taller than. We want to say that there is a 'real connection' between Fido and doghood, or between the

---

[10] For further discussion of this general point, see my 'Abstraction, Properties, and Immanent Realism', in T. Rockmore (ed.), *Proceedings of the Twentieth World Congress of Philosophy, Volume 2: Metaphysics* (Bowling Green, OH: Philosophy Documentation Center, 1999), pp. 195–205.

[11] Compare Kevin Mulligan, 'Relations—Through Thick and Thin', *Erkenntnis* 48 (1998), pp. 325–53, especially at pp. 344 ff.

[12] The phrase 'no addition of being' is used in a comparable way in various contexts by David Armstrong: see, for example, his *A World of States of Affairs*, p. 117. But I do not want my use of it to be tied rigidly to his.

roundness and the ball, which is absent between the objects of different heights. This, in part, is because in the former cases the relationships in question ground, or—as I earlier put it—constitute, *dependence* relations between the entities concerned. It is as if Fido and doghood, and the roundness and the ball, are 'made for each other', in a sense in which the objects of different heights are not. After all, each of the latter objects could have existed, just as it is, in the absence of the other—but Fido could not have existed in the absence of doghood, nor could the roundness have existed in the absence of the ball. So it won't do, apparently, *simply* to say that instantiation and characterization are 'internal' relations and as such 'no additions of being'.

3.5 FORM AND CONTENT IN ONTOLOGY

Instantiation and characterization clearly have something to do with being—they are, after all, *ontological relations*—and yet, for reasons given earlier, it seems that we cannot regard them as *beings*, or entities. The lesson, no doubt, is that there is more to the business of ontology than just saying *what there is*, or even what there could be. There is also the matter of saying *how* beings are, both in themselves and with respect to one another. One might have thought that this was just a matter of saying what the *properties* of beings are and what *relations* beings stand in to one another. But now we know that that can't be right, because this leaves out of account such matters as the *having* of properties—that is, what we have been calling 'characterization' (and, in another sense, 'exemplification'). We have established, for example, that the having of properties is not itself an element of being—a relational entity in which two other beings may stand to one another, the being that has a property and the property that is had by it.

But now we may begin to wonder why we needed to include properties and relations amongst the elements of being at all. Why not restrict ourselves to a one-category ontology of undifferentiated *entities*, or 'things', and say, with Quine, that the answer to the question 'What is there?' is simply given by the one word, 'Everything'?[13] On this view, the business of saying *what* there is ends very quickly, with the answer 'each and every thing', and all of the more interesting questions of ontology (if that is what we can still call it) are to do with *how* things are, understood in a way which involves the invocation of no further entities. For example, it might be said, one aspect of how things are is that some things are *dogs*. But that some things are dogs is not, on this view, a matter of there being things belonging to one category (that of objects) which instantiate something of another category (that of kinds): for all things, on this view, belong to the same category or—what amounts to the same thing—to no category at all. Does such a

---

[13] See W. V. Quine, 'On What There Is', in his *From a Logical Point of View*, 2nd edn (Cambridge, MA: Harvard University Press, 1961).

view really make sense, however? What, we may ask, entitles the adherent of such a view to say, in answer to the question of what there is, that there are *things*, in the plural, as opposed to just *something*? An adherent of the four-category ontology can certainly give a principled answer to this sort of question. Such an ontologist can point out, for instance, that *objects* can be many and distinct, in virtue of instantiating different *kinds*, which confer upon their instances their distinctive identity conditions. But the pure 'thing' ontologist, it seems, cannot claim that there is any objective fact of the matter as to whether reality consists of many things or just uncountable 'stuff'—the infamous 'amorphous lump'.[14] This way lies either radical antirealism or else some sort of Kantian or Schopenhauerian noumenalism.

What, then, is it about formal ontological relations that not only requires but permits us to regard them as having to do with *how* things are but not with *what* things there are? Surely, it is above all their *formal* character. But what does 'formal' mean in this context? The proper contrast here is between *form* and *content*. Such a contrast is, of course, also drawn in logic, between the logical form of a proposition and its non-logical content. But we are now concerned with a distinction of ontology, not of logic.[15] Ontology is in no sense a branch of logic. Beings, or entities, we may say, provide ontological content. But all beings also have ontological form. The ontological form of an entity is provided by its place in the system of categories, for it is in virtue of a being's category that it is suited or unsuited to combine in various ways with other beings of the same or different categories. An analogy is supplied by the chemical elements and their places in the periodic table, which determine how they may combine with other elements to form compounds. When beings do 'combine' in the ways to which they are suited by their ontological forms, these 'ways of combining' are the various formal ontological relations—instantiation, characterization and the rest. Such 'combinings' are 'no addition of being', just as there is, as it were, 'no addition of substance' when, in virtue of their valencies, an oxygen ion and two hydrogen ions combine to form a water molecule. Frege likewise seems to have had some sort of chemical model in mind in speaking as he did of 'saturation', with the similar purpose of denying that the combination of 'object' and 'concept' involved any further ingredient besides those two elements of being.[16]

If all of this still sounds intolerably obscure, let us focus for a moment on the most fundamental formal ontological relation of all, that of *identity*, which in a way provides a model for all of the rest. Some philosophers—for example,

---

[14] On the 'amorphous lump' conception of reality, see Michael Dummett, *Frege: Philosophy of Language*, 2nd edn (London: Duckworth, 1981), pp. 563 ff.

[15] For a comparable distinction between logical and ontological form, see Barry Smith, 'Logic, Form and Matter', *Proceedings of the Aristotelian Society*, Supplementary Volume LV (1981), pp. 47–63.

[16] See Gottlob Frege, 'On Concept and Object', in Peter Geach and Max Black (eds), *Translations from the Philosophical Writings of Gottlob Frege*, 2nd edn (Oxford: Blackwell, 1960). However, as we shall see in later chapters, the 'saturation' metaphor has its limitations.

Wittgenstein at one time—have denied that there is any such relation as identity.[17] And in a way they are right, if only for the wrong reasons. They point out, for instance, that it is trivial to say that each object is identical with itself, but blatantly false to say that any object is identical with any *other* object. So, it seems, all statements of identity must either lack content or else express necessary falsehoods. However, their lack of 'content' is just a reflection of the *formal* character of the identity relation. The self-identity of objects is, properly understood, a far from trivial matter. It is only in virtue of their self-identity that objects are countable and can constitute a plurality. This is not to say that self-identity is sufficient for countability, only that it is necessary. To be countable, objects must in addition instantiate kinds which confer upon them their distinctive identity *conditions*. Objects, that is to say, possess not only self-identity but also *unity*, which is what makes them individuatable.[18] But despite the indispensable ontological role that identity plays, identity is indeed not a 'relation'—not, that is to say, a relational universal, with relational modes as its particular instances. Self-identity, and hence identity, is, we might say, a metaphysically necessary condition of the existence of objects. That is no trivial matter. Without it, there could be nothing in the world. It is too fundamental, indeed, to be something *in* the world—an element of being—because it is that without which there could be no beings and so no world. And the same applies, I would say, to the other formal ontological relations.

3.6 COMPOSITION AND CONSTITUTION

Here, finally, is an appropriate place to say something more about those formal ontological relations that I have mentioned so far only in passing, namely, *composition* and *constitution*. Composition is the one-many relation between a (non-simple) whole and its proper parts—or, more accurately, between a (non-simple) whole and some of its proper parts, since the same whole may often have many different 'compositions'. For example, a copper statue is, at a given time, composed by some copper atoms, each copper atom being a proper part of the statue. But since each copper atom is in turn composed by various sub-atomic particles, the statue as a whole is also composed by the sub-atomic particles that compose each of the copper atoms. (Note here that the one-one relation of part to whole is just that relation in which one thing stands to another when the first thing helps to compose the second: that is, when the first thing is one of some things that compose the second thing. That is why I did not explicitly include this relation in my original list of indispensable formal ontological relations.)

---

[17] See Ludwig Wittgenstein, *Tractatus Logico-Philosophicus*, trans. D. F. Pears and B. F. McGuinness (London: Routledge and Kegan Paul, 1961), 5.5301, 5.5303.
[18] See further my *The Possibility of Metaphysics*, ch. 3.

Just as the identity conditions of an object are supplied by the kind that it instantiates, so the *composition conditions* of a composite object are supplied by the kind that it instantiates. Or, more accurately, the composition conditions of a composite object *at a certain level of composition* are supplied by the kind that it instantiates. For instance, the composition conditions of a copper statue at the level of its composition by copper atoms are supplied by the kind *copper statue* that it instantiates. In turn, the composition conditions of copper atoms at the level of their composition by sub-atomic particles are supplied by the kind *copper atom* that they instantiate. As a simple illustrative example, the composition conditions of a *lump of copper* may roughly be stated as follows: some copper atoms compose a lump of copper at a time t just in case (1) those copper atoms are fused together over a period of time to which t belongs and (2) during that period there are no other copper atoms with which any of them are fused. (Clause (2) here is required to reflect the fact that lumps of copper are 'maximal', that is, are not proper parts of larger such lumps.) It is easy to see that the composition conditions of a lump of copper differ significantly from those of a *copper statue*, since the latter must take account of the shape of the statue and allow for the possibility of a statue undergoing a change of composition over time at the level of its composition by copper atoms—a kind of change which is not admissible for lumps of copper. (Note here that the fact that objects of different kinds may have different composition conditions no more implies that 'composition' is ambiguous than does the fact that objects of different kinds may have different *identity* conditions imply that 'identity' is ambiguous.)

The foregoing example also provides us with an illustration of the relation of *constitution*. At any given time, a copper statue will be constituted by a certain lump of copper, although quite probably by different lumps at different times. The statue and the lump must be distinct (non-identical) entities, since they have different composition conditions and, correspondingly, different persistence conditions—which, of course, is why the same statue can be constituted by different lumps at different times.[19] *Constitution is not identity*. But, judging by this example, constitution would appear to be, like identity, a *one-one* relation, not—or, at least, not always—a one-many relation like composition.

But what exactly *is* the relation of constitution? And is there in fact just one relation that goes under this name? Earlier, I suggested that one formal ontological relation, such as rigid existential dependence, could be *constituted* by another—for example, by the relation of characterization in which a mode stands to the object of which it is a mode. What I was suggesting was that, when we say that a mode both *characterizes* and *depends existentially* upon the object of which it is a mode, we shouldn't *identify* these relations, but nor should we regard them as being quite unconnected. We want to say, perhaps, that it is *in virtue of* its

---

[19] See further my 'Coinciding Objects: In Defence of the "Standard Account"', *Analysis* 55 (1995), pp. 171–8, and also my *A Survey of Metaphysics*, ch. 4.

characterizing the particular object that it does that the mode depends existentially upon that object. The 'in virtue of' locution is typically apt wherever the relation of constitution is at issue. Thus, in the case of the copper statue and the lump of copper that constitutes it at any given time, we may say, for instance, that the statue has the shape and weight it does at that time *in virtue of* the shape and weight of the lump of copper. Constitution, it would seem, is the closest way in which two entities can be related while still remaining numerically distinct. Entities related by constitution are indiscernible in many respects, because the constituted entity inherits many of its features from the constituting entity—as the statue inherits its shape and weight from the lump of copper. But, being non-identical, they remain discernible in other respects—as the statue and the lump of copper do in respect of their histories and their potentialities. Of course, if we are really to regard the characterization relation between a mode and an object as *constituting* the dependence relation between them, and yet continue to insist that formal ontological relations like these are not *entities*, we shall have to allow that 'non-entities' as well as entities can be related by constitution. This would be problematic if constitution itself were a relational universal rather than a formal ontological relation—but not, I think, given that it is not.

4

Formal Ontology and Logical Syntax

4.1 TRADITIONAL FORMAL LOGIC VERSUS QUANTIFIED PREDICATE LOGIC

> [I]f objects and concepts are as different as Frege says they are, then it would seem entirely reasonable to hold that object-words and concept-words are as syntactically different as Frege makes them and the prohibition against predicating object-words is then also reasonable. I do not find much in this... [T]o anyone who finds an alternative to the atomicity thesis, the object-concept distinction loses its syntactical authority and becomes a metaphysical distinction of no great interest to the logician or the grammarian. The strength of the object-concept distinction is precisely the strength of Frege's logical syntax; the former is dependent on the latter and not vice versa.[1]

How far should logical syntax be expected to reflect ontology? And which, if either, is explanatorily prior to the other? The passage just quoted from Fred Sommers' ground-breaking book, *The Logic of Natural Language*, would suggest that he, at least, thinks that it is wrong-headed to attempt to make our logical syntax conform to our ontological preconceptions and, indeed, that some of these preconceptions may in any case be rooted in antecedently assumed syntactical distinctions which possess little merit in themselves. The 'atomicity thesis' of which he speaks in that passage is the assumption, central to the first-order quantified predicate logic of Frege and Russell, that the most basic form of elementary proposition is 'Fa', where 'a' is a proper name of a particular object and 'F' expresses a concept under which that object is said to fall (or, in another idiom, a property which it is said to exemplify). For such logicians, more complex propositions can then be formed by means of truth-functional operations or by quantifying into positions capable of being occupied by names of objects. Thus we obtain propositions of such forms as '$(\forall x)(Fx \rightarrow Gx)$' and '$(\exists x)(Fx \ \& \ Gx)$', which are standardly construed as being the logical forms of such natural-language sentences as 'All men are mortal' and 'Some philosophers are wise', respectively.

---

[1] Fred Sommers, *The Logic of Natural Language* (Oxford: Clarendon Press, 1982), p. 125.

Traditional formal logic, championed by Sommers himself, holds on the contrary that the logical forms of the latter two sentences are, respectively, 'Every *F* is *G*' and 'Some *F* is *G*' and takes these forms to be logically basic, along with such forms as 'No *F* is *G*' and 'Some *F* is non-*G*'. Such sentences are said to have subject–predicate form—that is, the form '*S* is *P*'—with terms symbolized as '*F*' and '*G*' taken as being capable of occupying either subject or predicate position. (Note here that such terms do not, in general, constitute *in their entirety* either the subject or the predicate of a sentence with subject–predicate form, according to traditional formal logic, because they need to be supplemented by logical particles such as 'every', 'some' and 'non'; I shall return to this important point in a moment.) By contrast, it is central to Frege–Russell logic that while 'atomic' propositions may legitimately be said to have subject–predicate form, in them only a *proper name* can occupy subject position. Correlatively, it is built into this conception of logical syntax that a proper name can never occupy *predicate position*, because according to it a 'predicate' just *is* what remains when one or more proper names are deleted from a sentence—and a proper name alone can never be all that remains when one or more other proper names are deleted from a sentence. As a consequence, Frege–Russell logic absorbs the 'is' of predication into the predicate, treating 'is wise' in 'Socrates is wise' on a par with 'thinks' in 'Socrates thinks'. In contrast, traditional formal logic does quite the reverse, reformulating the latter sentence so as to include explicitly a copula: 'Socrates is a thinking thing'.

Whereas for Frege and Russell elementary propositions are 'atomic', having such forms as '*Fa*' or (in the case of relational propositions) '*Rab*', and thus contain no logical particles, for Sommers all elementary propositions contain, in addition to non-logical terms, logical particles expressive of *quantity* or *quality* which are taken to be syncategorematic parts of the subjects and predicates, respectively, of those propositions (as was mentioned above). For example, 'some' is expressive of quantity and 'non' of quality in 'Some *F* is non-*G*'. The distinction between sentence negation and predicate negation is correspondingly crucial for Sommers—'It is not the case that some *F* is *G*' is not logically equivalent to 'Some *F* is non-*G*'—whereas it is entirely repudiated by Frege–Russell logic, for which sentence negation alone has any significance.

Frege and Russell disagree amongst themselves, of course, concerning the membership of the class of proper names—Russell restricting it to what he calls 'logically' proper names, while Frege is much more liberal, including even so-called definite descriptions. But this difference between them is minor in comparison with the doctrines that they hold in common and in opposition to those of traditional formal logic. One vital feature of Sommers' defence of traditional formal logic against the current Frege–Russell orthodoxy is his contention that proper names are themselves just 'terms', capable of occupying either subject or predicate position in elementary propositions as conceived by traditional formal logic—their only distinctive peculiarity being that, because such a name (unlike a so-called 'common name') is understood to be applicable to only one thing, it is a matter of indifference whether 'every' or 'some' is prefixed to such a name when it is in subject position. Thus, on Sommers' account, 'Socrates is a thinking thing' can indifferently be understood as saying either 'Every Socrates is a thinking thing' or 'Some Socrates is a thinking thing': a proper name has, as Sommers puts it, 'wild' quantity (by analogy with the notion of a 'wild card' in certain card games).[2]

Sommers holds that the Fregean notion of a distinctive 'is' of *identity* is just an artefact of Fregean logical syntax, without independent motivation.[3] The argument is that it is only because Frege insists that proper names (at least, when they are being used as such) cannot occupy predicate position in a subject-predicate sentence, and recognizes no logical role for 'is' as a copula, that he is compelled to postulate an 'is' of identity. (Frege himself famously cites the sentence 'Trieste is no Vienna' as a case in which 'Vienna' is *not* being used as a proper name.[4]) For consider the sentence 'Cicero is Tully', which contains two proper names. Because 'Tully' is the name of an object and so, according to Frege, cannot occupy predicate position, 'is Tully' cannot be treated by him as a simple monadic predicate on a par with 'is wise', in which the 'is' of predication has, allegedly, no genuine logical role. On the other hand, 'Cicero is Tully' cannot simply be treated as the juxtaposition of two proper names, with the logical form '*ab*', for no proposition can have that form. Hence the 'is' in 'Cicero is Tully' cannot be regarded as being, like the 'is' of predication, logically redundant: rather, it must express a *relational* concept, on a par with 'loves' in 'John loves Mary'. In accordance with the Fregean doctrine that a predicate is what remains when one or more proper names are deleted from a sentence, the 'is' in 'Cicero is Tully'—given that it must genuinely *remain* when both 'Cicero' and 'Tully' are deleted from that sentence—must be regarded as being quite as much a genuine predicate as the 'loves' in 'John loves Mary'.

For Sommers, as I say, Frege's need to recognize a distinctive 'is' of identity is merely an artefact of his doctrines concerning logical syntax. According to traditional formal logic as interpreted by Sommers, 'Cicero is Tully' has the form 'Every/some F is G', with 'Cicero' occupying subject position and 'Tully' occupying predicate position, while the 'is' is just the familiar logical copula which traditional formal logic regards as indispensable. So, although Sommers agrees with Frege that the 'is' in 'Cicero is Tully' is not logically redundant, he thinks that this is so because this 'is' is just the familiar copula which traditional formal logic always takes to be non-redundant: whereas Frege, who quite generally recognizes no genuine logical role for an 'is' of predication, is forced in this special case to recognize such a role for 'is', which thereby becomes elevated by him into a

---

[2] Sommers, *The Logic of Natural Language*, p. 15. [3] Ibid., ch. 6.
[4] See 'On Concept and Object', in *Translations from the Philosophical Writings of Gottlob Frege*, trans. P. Geach and M. Black, 2nd edn (Oxford: Blackwell, 1960), p. 50.

supposedly distinctive 'is' of identity. Sommers presses home his objection to this Fregean doctrine of identity by pointing out that Frege is obliged to postulate distinct and irreducible laws of identity to govern the logical behaviour of identity statements, whereas such laws appear to be immediate consequences of more general logical principles according to traditional formal logic. In particular, the laws of the reflexivity and substitutivity of identity (the latter being one construal of Leibniz's law) seem to be straightforwardly derivable: a fact of considerable significance, given that in modern quantified predicate logic the remaining laws of identity—the laws of symmetry and transitivity—are derivable from these two. For example, that 'Cicero is Cicero' is a logical truth seems just to be a special case of the logical truth of statements of the form 'Every F is F' and 'Some F is F' (assuming that 'Cicero' is not an empty name and remembering that 'Cicero' in subject position is supposed to have 'wild' quantity). And that 'Tully is wise' follows from 'Cicero is Tully' and 'Cicero is wise' seems just to be a special case of the fact that 'Some F is G' follows from 'Every H is F' and 'Some H is G'.[5]

While Sommers is critical of the claim promoted by many of Frege's latterday devotees, such as Geach and Dummett, that Fregean logical syntax constitutes a momentous and irreversible advance over what they see as the benighted syntactical doctrines of traditional formal logic, he is more charitable regarding the merits of modern first-order quantified predicate logic with identity than most modern logicians are regarding traditional formal logic. In fact, he is at pains to argue that traditional formal logic is at no expressive disadvantage with respect to quantified predicate logic, in that the formulas of each can be translated into corresponding formulas of the other.[6] It would seem that the main advantage he claims for traditional formal logic is that its syntax is much closer to that of natural language and consequently easier for people to manipulate and understand. It does not appear that he thinks that it makes much sense to ask which of these logical systems is 'correct', given that their formulas are intertranslatable—any more than, say, it makes sense to ask whether an axiomatic or a natural deduction formulation of quantified predicate logic is 'correct', even though each may have distinct practical advantages.

On the other hand, Sommers' criticisms of the Fregean doctrine of identity would suggest that he is in fact committed to a somewhat less ecumenical position. As we saw in the passage quoted from Sommers' book at the outset, it seems that Sommers favours—there, at least—the view that the logician's conception of syntactical structure should float free of ontological considerations.[7] However, it is difficult to see how one could accept Sommers' own complaint that Frege's doctrine of identity is merely an artefact of his views about logical syntax without

---

[5] See Sommers, *The Logic of Natural Language*, p. 129.
[6] See ibid., Appendix A. See also Fred Sommers and George Englebretsen, *An Invitation to Formal Reasoning: The Logic of Terms* (Aldershot: Ashgate, 2000).
[7] But see also the contrasting remarks at p. 305 of Sommers, *The Logic of Natural Language*.

rejecting, as a consequence, Frege's belief—which is clearly ontological in character—that there is such a relation as the relation of identity. More generally, it seems that we cannot entirely divorce Frege's syntactical doctrines from his ontological views. His distinction between object and concept may well be at least partly syntactical in origin, but it is indubitably ontological in import. And while we may very reasonably have doubts about the legitimacy of attempting to *found* ontological distinctions on syntactical ones—if that is indeed what Frege attempted to do—it is open to us to agree with certain of Frege's ontological insights and to concur with him on the relevance of those insights to questions of logical syntax.

4.2 THE TWO-CATEGORY ONTOLOGY OF FREGE–RUSSELL LOGIC

Suppose we agree with Frege in advocating something like his object/concept distinction, construed as capturing—or perhaps, rather, as conflating—aspects of two more familiar distinctions of traditional metaphysics: the distinction between *substance and property* and the distinction between *particular and universal*. (Frege himself no doubt thought that he was improving upon these older distinctions, by drawing on the mathematical distinction between function and argument, but we need not concur with him in this opinion.) Would it not then be reasonable to expect this distinction to be reflected in a perspicuous logical syntax? Logic is the science of reasoning and what we reason *about* are possible states of affairs and their relationships of mutual inclusion and exclusion.[8] We reason about them by representing them propositionally and reflecting on relations of mutual consistency and entailment amongst propositions. That being so, should we not aim to make logical syntax ontologically perspicuous by articulating or regimenting propositions in forms which reflect the constituent structure of the possible states of affairs that we are attempting to reason about? Suppose, for instance, that a certain possible state of affairs consists in some particular object's possessing some property, or exemplifying a certain universal. In that case, it seems that the state of affairs in question contains just two constituents—the particular object and the property or universal—which belong to fundamentally distinct ontological categories.[9] Would it not then be reasonable to represent this state of affairs by a proposition which likewise contains just two constituents of formally distinct types—by, indeed, an 'atomic' proposition of the classic Frege–Russell form 'Fa'?

---

[8] It would seem that Sommers himself is not unsympathetic to this sort of view: see his *The Logic of Natural Language*, ch. 8. Note, however, that I don't intend the remark in the text to indicate any commitment on my own part to an ontology which takes states of affairs to be the fundamental building blocks of reality, in the way that David Armstrong does (see following note).

[9] This, certainly, would be David Armstrong's verdict: see his *A World of States of Affairs* (Cambridge: Cambridge University Press, 1997).

Here, however, we may be beset by doubts of the sort famously raised by Frank Ramsey concerning the origins and legitimacy of the particular/universal distinction—doubts with which Sommers himself clearly has some sympathy.[10] Ramsey suggested that this supposed ontological distinction is a spurious one founded on a superficial understanding of the subject/predicate distinction. He pointed out that if we think of the proposition 'Socrates is wise' as somehow implicitly differentiating between Socrates conceived as being a 'particular' and wisdom conceived as being a 'universal' which characterizes that particular, then we should observe that we can restate the proposition in the equivalent form 'Wisdom is a characteristic of Socrates', in which 'wisdom' rather than 'Socrates' is now the subject of which something is said by means of the proposition's predicate.[11] (To emphasize the parallel between the two equivalent formulations, we could even contract the second to something like 'Wisdom is Socratic', without loss of significant content.) The suggestion then is that the entities picked out by the subject-terms of either of these equivalent sentences—Socrates and wisdom—do not really belong to distinct ontological categories in any intelligible sense. Each is just whatever it is and either can be truly and equivalently predicated of the other. My answer to such doubts is that it would indeed be mistaken to think that insight into the ontological distinction between particular and universal could be gained simply by reflection on the subject/predicate distinction—and even more seriously mistaken to suppose that the latter distinction could provide the foundation of the former. The lesson is that if the particular/universal distinction is a proper one to make, it must be one that is made on wholly ontological, not syntactical, grounds. More specifically, the only legitimate way to found the particular/universal distinction—or any other putative distinction between ontological categories—is, I consider, to provide a well-motivated account of how the existence and identity conditions of putative members of the one category differ quite generally from those of putative members of the other. And I believe that this can in fact be done in the case of the particular/universal distinction, as I have argued in earlier chapters and will explain again later. But first I need to provide a brief résumé, for the purposes of the present chapter, of the system of ontological categories that I am inclined to favour.

4.3 HOW TWO DISTINCTIONS GENERATE FOUR CATEGORIES

To begin with, let me return to a suggestion that I made in passing a little while ago, namely, that Frege's distinction between object and concept unhelpfully

---

[10] See Sommers, *The Logic of Natural Language*, pp. 41 ff.
[11] See F. P. Ramsey, 'Universals', in his *The Foundations of Mathematics and Other Essays* (London: Kegan Paul, 1931).

conflates two traditional metaphysical distinctions—the distinction between substance and property and the distinction between particular and universal. It is my view that we do indeed need to endorse both of these distinctions and that they should both be reflected in a perspicuous logical syntax. I consider that the two distinctions cut across one another, serving to generate *four* fundamental ontological categories: the category of particular substances, the category of particular properties (and relations), the category of substantial universals, and the category of property- (and relational) universals. This is precisely the *four-category ontology* (as I call it) that we find hinted at in the beginning of Aristotle's *Categories*, perhaps the most important single text in the history of ontology.[12] An example of a particular substance would be a certain particular dog, Fido. An example of a substantial universal would be the species or kind *dog* of which Fido is a particular instance. An example of a property-universal would be the colour—say, whiteness—which Fido has in common with other white things. And an example of a particular property would be the particular whiteness of Fido, which is necessarily unique to that particular dog and which is a particular instance of whiteness the property-universal. This is not the place for me to attempt to justify the four-category ontology—something that I do at length in other chapters of this book. I do, however, believe that we need to find room in our ontology for entities belonging to each of these categories and that in no case are the entities belonging to one of them wholly explicable in terms of, reducible to, or eliminable in favour of entities belonging to one or more of the others. This is what it means, in my usage, to say that these categories are *fundamental*. Moreover, I am not denying—any more than Aristotle did—that there are other ontological categories besides these four (for instance, it may be contended that there is a distinct ontological category of *events*): I am only committed to the view that any such further categories are not fundamental in my sense.

As well as recognizing these four fundamental ontological categories, we need to recognize two fundamental ontological relations in which entities belonging to these categories stand to one another. As the reader will know from previous chapters, I call these relations *instantiation* and *characterization*. Instantiation is the relation in which a particular instance of a universal stands to that universal: so, for example, it is the relation in which Fido stands to the kind *dog*, and it is the relation in which Fido's particular whiteness stands to whiteness the property-universal. Characterization is the relation in which properties stand to the substantial entities of which they are predicable. So, for example, Fido's particular whiteness characterizes—is a characteristic of—Fido. Analogously, whiteness the

---

[12] See further my *The Possibility of Metaphysics: Substance, Identity, and Time* (Oxford: Clarendon Press, 1998), pp. 203 ff. Here I should acknowledge that Sommers himself has important things to say about categories and is, of course, very sympathetic to Aristotle—see his *The Logic of Natural Language*, ch. 13. However, Sommers' approach to the theory of categories is very different from mine, according a central place to the notion of a 'category mistake' in the Rylean sense.

property-universal characterizes certain substantial universals or kinds, such as the kind *polar bear*. It does not characterize the kind *dog* as such, because not every kind of dog is white, just as not every kind of bear is white. What I have just said concerning characterization may be extended from the case of properties to the case of relations and the substantial entities which stand in those relations to one another. Thus, the particular relation of loving in which John stands to Mary characterizes John and Mary, taken in that order. Analogously, the relational universal of loving characterizes, perhaps, the human kinds *mother* and *child*, taken in that order. But I should emphasize that these putative examples are for illustrative purposes only, since not every relational expression should be assumed to denote a genuine relation, any more than every monadic predicate should be assumed to denote a genuine property. *That* we should include in our ontology properties and relations—both as particulars and as universals—I am now taking as given: but *which* properties and relations we should include is another question, to be settled by further discussion and argument, in which considerations of ontological economy and explanatory power will have an important and perhaps decisive role to play. This question is not one that concerns me at present, however.

I have spoken of instantiation and characterization as being fundamental ontological relations. But now I must qualify what I have just said, because in an important sense I do not think that they really are *relations*, that is, entities belonging either to the category of non-substantial universals or to the category of non-substantial particulars. In fact, I do not think that instantiation and characterization are *entities*—elements of being—at all, for reasons explained in Chapter 3 above. The fact that Fido instantiates the kind *dog* is not a *relational fact*, in the way that the fact that John loves Mary is. To put it another way, instantiation is not a *constituent* of the fact that Fido instantiates the kind *dog* in addition to its constituents Fido and the kind *dog*, in the way that their particular relation of loving is a constituent of the fact that John loves Mary in addition to its constituents John and Mary. As we saw in Chapter 3, one plausible way to account for this difference is to classify instantiation and characterization as so-called *internal* relations and to urge that such 'relations' are, in David Armstrong's useful phrase, 'no addition of being'.[13] An 'internal' relation, in the sense now at issue, is one in which its relata *must* stand, of logical or metaphysical necessity, given merely that they themselves exist and have the intrinsic properties or natures that they do. Perhaps the paradigm examples of internal relations in this sense are identity and distinctness. This, however, is not the place to dwell on this important issue any further. The crucial point is that we should *not* include instantiation and characterization amongst the relational entities belonging to either of our two fundamental categories of non-substantial beings. I like to register this point by calling instantiation and characterization purely *formal* relations—but we could

---

[13] See Armstrong, *A World of States of Affairs*, pp. 116 ff.

equally if more grandiloquently call them *transcendental* relations in order to register the same point.

4.4 THE ONTOLOGICAL SQUARE

Following my practice of previous chapters, we may conveniently represent the ontological scheme that I have just sketched by means of the diagram that I have proposed we call *the Ontological Square* (see Fig. 4.1).

One aspect of this scheme may seem puzzling, namely, that I speak of substantial particulars as being characterized by non-substantial particulars and as instantiating substantial universals—that is, *kinds*—but have said nothing explicitly so far in this chapter about any relationship between substantial particulars and *non-substantial* universals, that is, between an entity such as Fido, a particular dog, and an entity such as whiteness, the property-universal. However, it should be evident from the Ontological Square that there are in fact two different ways in which such entities may be related. On the one hand, a substantial particular may instantiate a kind which is characterized by a certain property-universal, while on the other a substantial particular may be characterized by a particular property which instantiates a certain property-universal. Thus, for example, on the one hand Fido instantiates the kind *dog* and the kind *dog* is characterized by the property-universal carnivorousness, while on the other hand Fido is characterized by a certain particular whiteness and this particular whiteness instantiates whiteness the property-universal. As I indicated in Chapter 2 and shall explain more fully in Chapter 8, my view is that it is this ontological distinction that underlies the semantic distinction between dispositional and categorical—or, as I prefer to call

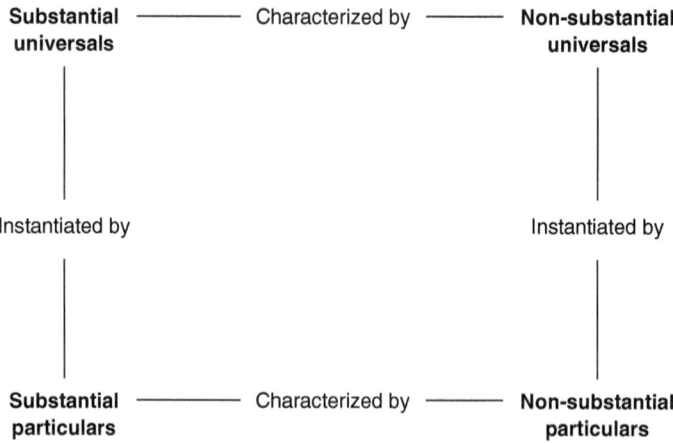

Figure 4.1. The Ontological Square (III)

the latter, *occurrent*—predicates. In effect, to say that a substantial particular is *disposed* to be *F* (or is 'potentially' *F*) is to imply that it is a thing of a *kind* which has *F*ness as a general feature, whereas to say that it is *occurrently F* (or is 'actually' *F*) is to imply that an instance of *F*ness is one of *its* particular features. Clearly, something may be disposed to be *F* even though it is not occurrently or actually *F*: for example, a crystal may be disposed to dissolve in water even though it is not actually dissolving in water. By my account, it has this disposition in virtue of being an instance of some chemical kind—such as the kind *sodium chloride crystal*—of which the property of dissolving in water is a characteristic. That the kind *sodium chloride crystal* has this characteristic is, however, nothing less than a *natural law*, most naturally expressible in English by the sentence 'Sodium chloride crystals dissolve in water', or, equivalently, 'Sodium chloride crystals are water-soluble'. So, in effect, my account of dispositionality may be summed up by saying that, according to it, a substantial particular, *a*, has a disposition to be *F* just in case there is a law connecting *F*ness with the kind of thing that *a* is.[14]

I do not expect this ontological scheme to seem entirely compelling simply on the basis of the very sketchy account of it that I have provided here, but it is not vital for my present purpose that I should be able to convince the reader of its correctness. My current purpose is merely to persuade the reader that it is legitimate and indeed desirable to tailor one's theory of logical syntax to one's ontological convictions. For this purpose, I need nothing more than the reader's concurrence that the four-category ontology as I have sketched it is at least coherent and worthy of consideration. From this point on, then, I shall take the ontology to be as I have described it and attempt to show how our account of logical syntax needs to be adapted in order to accommodate that ontology in a perspicuous fashion. But before doing so I must redeem a promise made earlier, to indicate how the categorial distinctions that I have invoked may be explicated in terms of the distinctive existence and identity conditions of the members of the categories concerned.

Very briefly, then, what I have to say on this matter is as follows—a fuller account is provided in other chapters. Non-substantial particulars depend for their existence and identity upon the substantial particulars which they characterize. For example, Fido's particular whiteness exists only because Fido does and is distinguished from other particular whitenesses precisely in being *Fido's*. In contrast, Fido depends neither for his existence nor for his identity upon his particular whiteness, since he could exist without it (if, for example, he were to change colour). We may summarize this sort of ontological asymmetry between substantial and non-substantial particulars by saying that the former are *independent* and the latter *dependent* particulars. As for universals, whether substantial or non-substantial, they are dependent entities in another sense, at least according to

---

[14] See further my *Kinds of Being: A Study of Individuation, Identity and the Logic of Sortal Terms* (Oxford: Blackwell, 1989), ch. 8 and, in the present book, Chapter 8 below.

the sort of immanent or 'Aristotelian' realism concerning universals that I favour, for they are 'generically'—or, in the terminology of Chapter 3, *non-rigidly*—existentially dependent upon their particular instances. That is to say, according to this immanent realist view, a universal can exist only if it has *some* particular instances: there are and can be no uninstantiated universals. But a universal is not dependent for its *identity* upon its particular instances: the very same universal could have had different particular instances from those that it actually has. Finally, we need to say something about the distinction between substantial and non-substantial universals. The plausible thing to say here is this. First, non-substantial universals depend for their existence but not for their identity upon substantial universals. For instance, whiteness the property-universal plausibly would not exist if no *kind* of thing had the characteristic of being white, but whiteness is not the very property that it is in virtue of the kinds of things that it characterizes—for it could have characterized quite other kinds of things. Second, substantial universals depend for their identity—and hence also for their existence—upon at least some non-substantial universals. Thus, the kind *dog* depends for its identity upon a number of property-universals—its 'essential' characteristics—which include, for example, the properties of warmbloodedness and carnivorousness. I would be the first to acknowledge that this account of the matter in hand involves a considerable amount of oversimplification, but I leave to later chapters the task of providing a fuller treatment of some of the issues involved.

4.5 SORTAL LOGIC AND THE FOUR-CATEGORY ONTOLOGY

Now let us return to the question of logical syntax. The first thing to observe here is that modern first-order predicate logic with identity—the logic of Frege and Russell—is patently inadequate to the representational requirements of the four-category ontology. In effect, this logic is tailored to the needs of a *two*-category ontology, which includes only the category of substantial particulars (in Fregean terms, 'objects') and the category of non-substantial universals (in Fregean terms, 'concepts'). This limitation imposes a number of syntactical distortions from the perspective of the four-category ontology. One such distortion is the inability of Frege–Russell logic to distinguish syntactically between a proposition affirming that a substantial particular instantiates a certain substantial kind and a proposition affirming that a substantial particular exemplifies a certain property—between, for instance, 'Fido is a dog' and 'Fido is white'. Both propositions are represented in Frege–Russell logic as having the form '*Fa*'. Another such distortion is the inability of Frege–Russell logic to distinguish syntactically between a proposition affirming that a certain substantial kind is characterized by a certain property and a proposition affirming that some or all of the substantial particulars which instantiate that kind exemplify that property—between, for instance, 'Polar bears are white' (or 'The polar bear is white') and either 'Some polar bears are white' or 'All polar bears are white'. Both propositions are represented in Frege–Russell logic as having either the form '$(\exists x)(Fx \,\&\, Gx)$' or the form '$(\forall x)(Fx \to Gx)$'. A third example is the inability of Frege–Russell logic to distinguish syntactically between a proposition affirming that a substantial particular is disposed to be F and a proposition affirming that a substantial particular is occurrently F— between, for instance, 'Fido growls' or 'Fido eats meat' and 'Fido is growling' or 'Fido is eating meat'. Both propositions are again represented in Frege–Russell logic as having the form 'Fa'.

All of these distortions can be eliminated by means of some simple syntactical amendments to Frege–Russell logic.[15] First, we can include two different types of individual constants and variables, one to name and range over substantial particulars and one to name and range over substantial universals—$a, b, c, \ldots x, y, z$ on the one hand and $\alpha, \beta, \gamma, \ldots \phi, \chi, \xi$ on the other. Next, we can include, in addition to the identity sign, '=', a sign for instantiation, '/'. Finally, we can adopt the convention that when individual constants and variables are written *after* a predicate or relation symbol *occurrent* predication is intended, whereas when they are written *before* a predicate or relation symbol *dispositional* predication is intended. The modifications can be illustrated using the examples deployed earlier. First, then, 'Fido is a dog' will be symbolized in the form 'a/β', whereas 'Fido is white' (assuming that the predication is intended to be occurrent) will be symbolized in the form 'Fa'. Next, 'Polar bears are white' will be symbolized in the form 'γF' (with the predication taken to be dispositional in intent), whereas 'Some polar bears are white' and 'All polar bears are white' will be symbolized in the forms, respectively, '$(\exists x)(x/\gamma \,\&\, xF)$' and '$(\forall x)(x/\gamma \to xF)$' (again assuming that the predication is intended to be dispositional).[16] Finally, 'Fido growls', taking the predication here to be dispositional in intent, will be symbolized in the form 'aG', whereas 'Fido is growling' will be symbolized in the form 'Ga'. Moreover, if the account of dispositionality advanced earlier is correct, our extended logical syntax allows us to represent that fact by affirming the logical equivalence of a proposition of the form 'aG' with one of the form '$(\exists \phi)(a/\phi \,\&\, \phi F)$'.[17]

It will be noted that the extensions we have so far made to the syntax of Frege–Russell logic do not yet fully accommodate the representational requirements of the four-category ontology, because we have as yet made no provision for quantification over *non-substantial entities*. However, this limitation could easily be overcome in obvious ways. The important point is to see that one's ontology really does impose some constraints on what will serve as a perspicuous logical syntax, in which necessary connections internal to that ontology may be reflected adequately in formal relations between propositions.

[15] For more details, see my *Kinds of Being*, ch. 9.
[16] That even colour predicates exhibit the dispositional/occurrent distinction is argued for in my *Kinds of Being*, ch. 8; see also Chapter 8 below.   [17] Compare my *Kinds of Being*, p. 170.

Elsewhere, I have called the kind of logic for which this extended logical syntax is designed *sortal* logic.[18] Sortal logic is 'first-order' logic in the sense that it does not, as we have just noted, involve any quantification over *properties or relations*. But a logician of the Frege–Russell school could not see it in this light, because sortal logic does involve reference to and quantification over *substantial universals*, which such a logician—being wedded to a two-category ontology—is unwilling to distinguish from properties. On the other hand, from the point of view of the traditional formal logician, it looks as though sortal logic agrees with his system—in opposition to modern predicate logic—in allowing 'general terms' to occupy both subject and predicate position in elementary sentences, because it recognizes sentences both of the form 'a/β' and of the form 'βF'—for example, both 'Fido is a dog' and 'A dog eats meat' (understanding the latter as a variant form of 'Dogs eat meat'). However, this is in fact a somewhat misleading way of representing the situation, because sortal logic does not really treat 'Fido is a dog' as a subject–predicate sentence, on a par with, say, 'Fido growls' or 'Fido eats meat'. According to sortal logic, the 'is' in 'Fido is a dog' is not the 'is' of predication, which, in common with Frege–Russell logic, it regards as being logically redundant. In fact, sortal logic goes further than Frege–Russell logic in including *two* logical relation symbols, rather than just one: the sign of instantiation as well as the sign of identity, with the former appearing in 'Fido is a dog' and the latter appearing in, say, 'Fido is Rover'. (However, it is technically feasible in sortal logic to define identity in terms of instantiation, if it is allowed that every substantial entity trivially instantiates *itself*: for then we can say that $a=b=_{df} a/b \ \& \ b/a$ and $\alpha=\beta=_{df} \alpha/\beta \ \& \ \beta/\alpha$.[19]) Finally—although this is something that I have not alluded to until now—it seems that sortal logic, unlike Frege–Russell logic but once again like traditional formal logic, has need of a distinction between predicate negation and sentence negation. This arises from its recognition of a distinction between dispositional and occurrent predication, because it seems clear that there is a difference between affirming that a is disposed to be non-F and denying that a is disposed to be F—between, for example, saying 'Fido *doesn't* growl' and saying '*It's not the case that* Fido growls'.[20]

No doubt Sommers would object to my recognition of a distinctive 'is' of instantiation quite as strongly as he objects to Frege's recognition of a distinctive 'is' of identity.[21] In reply I would say, for reasons which should by now be clear, that without recognizing such a distinctive 'is' we cannot perspicuously reflect, in our propositional representation of many of the possible states of affairs that we

[18] See my *Kinds of Being*, ch. 9.
[19] See ibid., pp. 39–40 and pp. 183–4.
[20] See ibid., pp. 191 ff, where it is argued more generally that we need to distinguish between compound sentences and compound predicates.
[21] Such an objection is indeed raised by George Englebretsen in his thoughtful review of my *Kinds of Being*: see *Iyyun, The Jerusalem Philosophical Quarterly* 40 (1991), pp. 100–5. In that review, Englebretsen takes me to task for 'put[ting] the ontological cart before the logical horse' (p. 103). I accept the charge, but disagree about which should be described as the horse and which as the cart.

need to reason about, the constituent structure of those states of affairs—on the assumption, of course, that the four-category ontology is correct. It is my opinion, in opposition it seems to Sommers, that the theory of logical syntax should not and indeed cannot be ontologically neutral. It may be that logic should be free of existential commitments, but it cannot, I think, be profitably regarded as metaphysically innocent. Sommers' critique of modern predicate logic and the hegemony of the Frege–Russell legacy is entirely to be welcomed for helping to open our eyes to the distortions which that legacy has imposed upon prevailing conceptions of logical syntax. But rather than seeking the remedy in an attempt to eschew ontological considerations altogether where questions of logical syntax are concerned, I prefer to look for the remedy in a reformation of ontology and a renovation of logical syntax in alignment with it.

PART II
OBJECTS AND PROPERTIES

5

The Concept of an Object in Formal Ontology

5.1 OBJECTS, THINGS, AND ENTITIES

Many philosophers use the term 'object' in a rather imprecise way. Often they treat it as being interchangeable with the more colloquial term 'thing', which itself has a variety of more or less general senses. In one sense, things are contrasted with, for instance, events and processes and are therefore said to exist but not to happen or occur and to have spatial but not temporal parts. In another sense, everything whatever that there is or could be, including events, processes, states, properties, numbers and propositions, is said to be a 'thing' or 'object'—and this, indeed, is implicit in how the word 'thing' is understood when it is combined with quantifying adjectives like 'any' and 'some' to form the unitary general quantifiers 'anything', 'something', 'everything', and 'nothing'. This is why it is trivially true that every *thing* is a thing, in whatever sense 'thing' is understood, but not trivially true that *everything* is a thing, except when 'thing' is understood in the most general possible sense. 'Thing' or 'object' in its most general possible sense is, then, equivalent with the perfectly general and entirely topic-neutral terms 'entity' and 'item'.

In formal ontology, we need to be more precise in our use of the term 'object' than philosophers generally are. Formal ontology is that branch of analytical metaphysics whose business it is to identify ontological categories and the formal ontological relations that characteristically obtain between the members of different categories. In this context, the term 'thing' is best avoided because, being in widespread colloquial use, we can hardly hope to recruit it satisfactorily for precise technical purposes. If a perfectly general term is wanted that is applicable to members of any ontological category whatever, we do best to employ what is already a semi-technical term, 'entity'. That choice having been made, we should then use the term 'object' in a more restrictive sense, to apply to some but not all possible entities. Such a decision is not merely arbitrary or stipulative, however, because many philosophers already do use the term in a deliberately contrastive way. Formal ontologists can usefully exploit this existing practice and give it a more rigorous foundation in a theory of categories and formal ontological relations.

5.2 OBJECTS AND PROPERTIES

It would be consonant with the existing philosophical use of the term 'object' in its more restrictive sense to build into the formal concept of an object a contrast between *objects* on the one hand and *properties and relations* on the other. Objects are thus conceived to be entities that are by their very nature property-bearers and entities that do or can stand in various relations to one another. The distinction is to some extent reflected in language in terms of the syntactical distinction between subject and predicate, or between singular noun phrases on the one hand and adjectival or verb phrases on the other. But formal ontologists must constantly be on their guard against simplistic attempts to read ontological distinctions out of syntactical ones. Ontology, properly understood, is not merely the shadow of syntax. Syntax has no doubt evolved in a way that is partially sensitive to ontological distinctions, but is influenced by many other factors which make it an unreliable guide to ontology.

To insist, in the way I suggest we do, on a contrast between objects on the one hand and properties and relations on the other is to imply, of course, that it would be a mistake—in fact, a *category* mistake—to speak of properties and relations as themselves being objects. Of course, philosophers of a nominalist persuasion may well want to deny that properties and relations are objects because they want to deny their existence altogether. For the most part, what such nominalists are denying is the existence of properties and relations conceived as *universals*, since many self-styled nominalists are happy, these days at least, to acknowledge the existence of properties and relations conceived as particulars, namely, as so-called *tropes*. However, trope theorists are generally content to deny that tropes are *objects*, holding as they typically do that objects themselves are 'bundles' of tropes.

One thing that immediately emerges from these observations is that it would be a serious error simply to conflate the object/property distinction with the particular/universal distinction, since amongst those philosophers who believe in the existence of properties and relations it is disputed whether these entities are universals or particulars. However, I shall return to the particular/universal distinction later and concentrate for the time being on the object/property distinction. Properties and relations, I have said, are to be contrasted with objects and hence are not to be spoken of as being objects themselves. But now an immediate difficulty seems to arise, which is that it appears that properties and relations can themselves be said to possess properties and to stand in relations. This may lead us to suppose that the object/property distinction cannot, after all, be an absolute but at most only a relative one, so that any entity is an 'object' with respect to any other entity which it possesses as a 'property'. This, however, is a suggestion that I want to repudiate very strongly. It is a suggestion which is largely rooted, I think, in the overreliance on syntax as a guide to ontology.

We have what might be described as being a *canonical* way of referring to properties through the nominalization of predicates, typically with the help of a functorial expression like 'the property of . . .'. Thus, starting with the simple subject–predicate sentence 'The flower is red' and deleting the subject term, we are left with the predicate '. . . is red'. The copula in this predicate, 'is', may then be transmuted into the gerundial form 'being', and the resulting phrase can then either be prefixed with the above-mentioned functorial expression to give the noun phrase 'the property of being red', or it can be used to much the same effect as a noun phrase in its own right, 'being red'. It is, of course, fatal to suppose that such nominalizations of predicates always succeed in referring to something, since this leads us directly into the property version of Russell's paradox. We know that, on pain of contradiction, there can be no such thing as the property of being non-self-exemplifying, for instance, even though '. . . is non-self-exemplifying' is a perfectly meaningful predicate.[1] We should bear this caution closely in mind in considering whether there is really good reason to think of properties as being themselves property-bearers.

Consider the following sentence, which seems to express an uncontentious truth: 'The property of being red is a colour-property'. This should not be conflated with the sentence 'The property of being red is a colour', whose truth is at least questionable. 'Red is a colour' is plausibly a true sentence, but 'red' as it occurs in this sentence is not uncontentiously co-referential with 'the property of being red', even if we accept that the latter genuinely refers to something. Exactly how these three sentences are related to one another semantically is a difficult question, one lesson of which is, again, that we must not incautiously attempt to read off ontology from the surface syntax of natural language. Even if we assume that there is such a property as the property of being red and that 'The property of being red is a colour-property' is a true sentence, it is another matter to conclude that the property of being red possesses *the property of being a colour-property*, because we have no automatic right to assume that the noun phrase 'the property of being a colour-property' refers to anything at all.

Here it may be inquired what the *truthmaker* of the sentence 'The property of being red is a colour-property' is, if not the fact that one property—the property of being red—possesses another, higher-order property. However, allegiance to the truthmaker principle should not lead one to assume that truths stand in some sort of simple correspondence relation to truthmakers. For instance, one possibility is that what makes it the case that a certain property is a *colour*-property is the fact that objects bearing the property are, in virtue of bearing it, coloured in one way or another. Such an explanation of the truth of the sentence in question makes no obvious reference to any sort of higher-order property and so provides no reason to

---

[1] I discuss this more fully in my 'Abstraction, Properties, and Immanent Realism', *Proceedings of the Twentieth World Congress of Philosophy, Volume 2: Metaphysics*, ed. T. Rockmore (Bowling Green, OH: Philosophy Documentation Center, 1999), pp. 195–205.

suppose that properties are themselves bearers of properties as well as being borne by certain objects. I am not convinced, in fact, that there are any good reasons at all for including so-called higher-order properties in our ontology—and not because I have any nominalistic inclination to exclude properties altogether. That being so, the spectre of higher-order properties poses no threat to the claim that the object/property distinction is an absolute rather than a relative one.

On the other hand, I do not want to leave the absoluteness of the object/property distinction hostage to a satisfactory resolution of the question of whether higher-order properties exist. There is, however, another way in which we can secure its absoluteness, and this is by building into the concept of an object the requirement that an object is not just a property-bearer, but is also not itself capable of being borne or possessed in the sense that properties are. In other words, if we allow there to be, at least in principle, a hierarchy of property-bearing entities of ascending orders, then objects are to be characterized as being those entities that occupy the lowest level of this hierarchy, or as being of *order zero*. First-order properties are then properties of objects, second-order properties are properties of properties of objects, and so on. This way of conceiving of objects is Aristotelian in spirit, because it represents objects as having at least one hallmark of Aristotelian individual substances. However, not all objects, thus conceived, qualify as Aristotelian individual substances, because the latter are additionally conceived as being, in a certain sense, ontologically independent entities. Spelling out the precise sense of ontological independence involved here is a complex matter, which I shall not go into at present.[2] Suffice it to say that an object such as a pile of rocks would not qualify as an Aristotelian individual substance, because the pile depends for its existence and identity on the rocks which compose it and cannot survive the destruction or replacement of any one of those rocks.

5.3 RAMSEY'S PROBLEM

This is perhaps a good place at which to recall a type of objection to the object/property distinction which is often associated with the name of Frank Ramsey, although he raised the objection with regard to the particular/universal distinction which, as I have already pointed out, should not be conflated with the object/property distinction.[3] Essentially, Ramsey's point (transmuted into a point about the latter distinction) is that we gain no real purchase on what distinguishes objects from properties by saying that the former 'possess' the latter whereas the latter 'are possessed by' the former, because any statement attributing a property to

---

[2] I do so more fully in my *The Possibility of Metaphysics: Substance, Identity, and Time* (Oxford: Clarendon Press, 1998), ch. 6.

[3] See F. P. Ramsey, 'Universals', in his *The Foundations of Mathematics and other Logical Essays* (London: Kegan Paul, 1931). What Ramsey actually says is best interpreted as making a point about the object/property distinction.

an object can be recast as a statement in which what was formerly attributed becomes that to which something is attributed, and vice versa. For example, 'Socrates possesses the property of being wise' can be recast as 'Wisdom possesses the property of being Socratic', where we stipulate that the property of being Socratic is possessed by something just in case it would be true to say that Socrates possesses that thing. So, specifically, wisdom possesses the property of being Socratic, because it is true to say that Socrates possesses wisdom. The lesson we are supposed to draw from this kind of consideration is that 'Socrates' no more exclusively refers to an object, as opposed to a property, than 'Wisdom' exclusively refers to a property, as opposed to an object—and, more generally, that the object/property distinction is merely an artefact of language without serious ontological significance.

In my judgement, this kind of reasoning leads to a conclusion that is hostile to the ambitions of formal ontology only because such a hostility is built into it from the start. Let it be granted that 'Socrates possesses the property of being wise' can be translated as 'Wisdom possesses the property of being Socratic', when the predicate '... is Socratic' is given the special sense required. It still doesn't follow that we have been given any reason whatever to suppose that neither Socrates nor wisdom is distinctively an object as opposed to a property. For we must recall that we need to exercise extreme caution before assuming that an apparently property-denoting expression refers to anything at all. We have been given no reason to suppose that there is any such entity as the property of being Socratic. The predicate '... is Socratic' may certainly be assigned a meaning such that 'Wisdom is Socratic' turns out to be true just in case 'Socrates is wise' is true. But the latter, we may maintain, is true just in case the object Socrates possesses the property of being wise, or wisdom, and consequently that this is also what makes the sentence 'Wisdom is Socratic' true. We certainly needn't concede that we have no better reason to say this than to say that what makes these sentences true is the fact that the 'object' wisdom possesses the 'property' of being Socratic. For, as I say, we have been given no reason to suppose that there is any such entity as the property of being Socratic. In short, it is because the Ramsey-style objection already lays excessive emphasis on the mere surface syntax of language that it comes to its conclusion that the object/property distinction is nothing more than a superficial reflection of an essentially arbitrary syntactical distinction.

Even so, we shouldn't dismiss the Ramsey-style objection without further thought, because it does teach us something important, and this is not merely that the proper way to do ontology is not by appealing to syntax. The more interesting lesson is that it is not enough, in order to characterize the object/property distinction effectively, simply to stipulate that objects are entities of 'order zero' in the putative hierarchy of property-bearing entities. For merely saying this does not serve to distinguish between objects and certain possible *highest* order entities in the hierarchy. If the hierarchy contains such entities, then it is a hierarchy which includes both a lowest and highest level, such that entities at any given level can only stand in the asymmetrical bearer/borne relation to entities at immediately adjacent levels. But this fact in itself does not suffice to tell us *which* of the two extremum levels is the 'lowest' and *which* the 'highest', and so which entities are 'objects' and which are supreme 'properties'. In short, we need to build more into the object/property distinction than is simply given by the structure of the hierarchy into which both objects and properties fit.

Fortunately, more is indeed available for this purpose and, once again, we can turn to Aristotle for guidance. I remarked earlier that not all objects are Aristotelian individual substances, because not all of them have the right kind of ontological independence. A pile of rocks is not an individual substance, because it depends for its existence and identity on the rocks that compose it. Now, for Aristotelians, properties too lack the ontological independence of individual substances, but in a rather different way from that in which objects like piles of rocks do. Properties depend ontologically on the objects that bear them—and this is true whether or not those objects are individual substances. For Aristotelians, properties need bearers in order to exist and there cannot be, as Platonists suppose, unborne properties. Of course, objects also need properties in order to exist, because an object, as we are proposing to understand the term, is essentially a property-bearer. But if properties need objects to bear them and objects need to bear properties, it may seem that there is a perfect symmetry of ontological dependence between objects and properties which will defeat any attempt to distinguish between them in terms of such dependence. Here, however, we need to appreciate that there are many different varieties of ontological dependence and that the way or ways in which properties depend upon objects differ from that in which objects depend on properties. The property of being electrically charged would not exist if there were no electrically charged objects. But many electrically charged objects are such that, although they must of course have *some* properties in order to exist, they need not have the property of being electrically charged in order to exist, because they have this property only contingently. This is not the place for me to try to spell out in detail the asymmetries of ontological dependence between objects and properties, not least because the issue turns importantly on whether properties are conceived to be universals or particulars (tropes)—a question that I shall address in detail in Chapter 6. I merely want to point out that a promising strategy for metaphysicians with Aristotelian sympathies, when confronted with the problem of distinguishing between objects and properties, is to appeal to the distinctive way or ways in which properties depend ontologically upon the objects that bear them.[4] I shall return to this task in Chapter 7, where I shall explore more fully the implications of Ramsey's sceptical arguments.

---

[4] See again my *The Possibility of Metaphysics*, ch. 6.

5.4 THE INDIVIDUALITY OF OBJECTS

So far, then, we have built into the formal concept of an object the idea that objects are property-bearers but are not themselves borne, in the way that properties are, by other entities and we have also built into it the idea that properties are, in a characteristic fashion, ontologically dependent upon objects. Objects, we may say in Aristotelian vein, are ontologically prior to properties, or occupy a more fundamental place in the scheme of being. However, this is still not enough, in my view, to render the concept of an object sufficiently precise for the purposes of formal ontology. I want to build in additionally the idea that objects have *determinate identity conditions* and are, in virtue of their unity, *countable entities*. The reason for doing so is to distinguish objects from other actual or possible property-bearers of order zero which lack these formal characteristics. In effect, what I am thus proposing to build into the concept of an object is the idea of *individuality*.[5] We can best see how this applies by contrasting paradigm examples of objects with certain property-bearing entities of other types.

For example, then, quantum 'particles', such as electrons and photons, are property-bearers, but apparently lack fully determinate identity conditions. Thus, there are two electrons orbiting the nucleus of a helium atom and both of these electrons have, for instance, the property of being negatively charged. But it does not appear that there is, so to speak, any 'fact of the matter' as to which of these electrons is which. The electrons are two in number but it seems that we cannot, even in principle, attach a name or label distinctively to either one or refer demonstratively to either one in distinction to the other. This is despite the fact that the electrons are evidently *countable* entities, being precisely two in number. It is true that we can say, correctly, that one of the electrons has a 'spin' in one direction and the other has a 'spin' in the opposite direction: this is in fact required by the Pauli exclusion principle, which precludes two or more electrons being in exactly the same quantum state. However, we cannot say *which* of the electrons has a spin in which of the two directions. And this is not a mere reflection of our ignorance of the exact situation in which the electrons find themselves, but a reflection of the very nature of their situation as quantum theory represents it as being. Electrons, then, are not *objects*, in the precise sense which I recommend for the use of that term.[6]

For my next example of non-objects I turn to a hypothetical case, that of quantities of homogeneous or 'homoeomerous' matter, sometimes affectionately called 'atomless gunk' (David Lewis's term). Such quantities of matter are property-bearers—for instance, they possess the property of being mutually

---

[5] I say more about this in my 'Individuation', in M. J. Loux and D. W. Zimmerman (eds), *The Oxford Handbook of Metaphysics* (Oxford: Oxford University Press, 2003) and in my 'Identity, Individuality, and Unity', *Philosophy* 78 (2003), pp. 321–36.

[6] For further discussion of this, see my *The Possibility of Metaphysics*, pp. 62 ff.

impenetrable—but they lack individuality, albeit for a different reason from that for which quantum particles do. For, while they would seem to have fully determinate identity conditions, they lack the intrinsic unity and hence countability requisite for individuality. A particular quantity of matter may be gathered together into a compact mass, or it may be scattered haphazardly across the universe. In principle, it might even be smeared out thinly but continuously throughout the entirety of space. It qualifies in itself neither as a one nor as a many, that is, neither as a single object nor as a plurality of objects. This is why such quantities of matter are not countables. Of course, if such a quantity is gathered together into a compact mass, say in the shape of a ball, then it *composes* a single individual, namely, a certain ball of matter. We can count distinct, non-overlapping balls of matter, but what we are thus counting are not *quantities* of matter as such, but objects composed by certain quantities of matter.[7] We see, then, that there are two independent ways in which property-bearing entities may lack individuality and so fail to qualify as objects in my recommended sense: they may lack it either by lacking fully determinate identity conditions or they may lack it by lacking intrinsic unity and hence countability.

5.5 PARTICULARS AND UNIVERSALS

Let us now take stock of where we stand. Objects, I have said, are property-bearing entities of order zero, having a certain kind of ontological priority over property-bearing entities of higher orders (if such there be), and are also individuals, having determinate identity and countability. This still leaves us with a number of important issues to discuss concerning the place of objects in an adequate ontological scheme. First of all, we need to return to the distinction between particulars and universals and see how the concept of an object relates to that distinction. It might be supposed that universals, if they exist, are obviously either properties or relations and as such cannot qualify as objects. As we shall shortly see, matters are not as simple as this. But before we can do that, we must attempt to characterize the particular/universal distinction itself. Some philosophers attempt to characterize it in spatiotemporal terms, by saying that universals are, whereas particulars are not, 'multiply locatable' entities. The idea is that a universal, such as redness or the property of being red, exists in its entirety in every spatial location that is occupied by a red object—it is, as they say, 'wholly present' in each of those locations and nowhere else. However, not only is the notion of multiple locatability problematic in itself, but also it can obviously have no application to those universals, if such there be, that are exemplified by *abstract* objects, which do not—according to many philosophers, at least—exist in space

---

[7] For further discussion, see my *The Possibility of Metaphysics*, pp. 72 ff and my 'Identity, Individuality, and Unity'.

and time at all.[8] Consider, for example, the property of being an even prime, which is possessed exclusively by the number two. Since the number two plausibly does not exist in any spatial location whatever, neither can the property of being an even prime.

A more promising way to characterize the particular/universal distinction is in terms of the *instantiation* relation which obtains between a universal and its particular instances. In effect, we define a universal as being an entity which does or at least (if we are Platonists) *can* have instances and a particular as being an entity which *cannot*. Difficulties are created for this definition if it is supposed that there may exist universals which, because they embody some sort of contradiction or inconsistency, cannot have instances—such as, perhaps, the property of being a round square cupola. However, there are various strategies for getting around such difficulties and so I shall dwell on them no further and assume that something like our proposed definition is satisfactory.[9] But now it may seem that in adopting this definition we are simply revisiting our earlier discussion of the object/property distinction. In other words, it may seem that in attempting to characterize the particular/universal distinction in terms of the *instantiation* relation between particulars and universals, we are just replicating our earlier attempt to characterize the object/property distinction in terms of the *possession* relation between objects and properties. But that is not so, because it is vitally important in formal ontology to distinguish clearly between these two formal relations, instantiation and possession—or, as I often prefer to call the latter, *characterization*. (In an older terminology, it is called *inherence*—or, to be more precise, 'inherence' or 'characterization' is the name for the inverse of the possession relation.) Particulars *instantiate* (that is, are instances of) universals but do not *possess* universals. Objects *possess* properties but do not *instantiate* properties.

We are now in a position to see that not all objects, at least as I have so far defined them, are necessarily particulars. There can, in principle, be entities which are universals, because they have particular instances, but which are also objects, in that they are individuals possessing properties and standing in relations. Are there any such entities? In my view, there certainly are, the paradigm examples being what I call *kinds* of particular objects.[10] Consider, for example, a particular dog, Fido, and the canine kind to which it belongs. Fido is, perhaps, a poodle and, certainly, Fido is a dog and so also a mammal. Fido, then, instantiates the kinds *poodle*, *dog* and *mammal*. These kinds are characterized by distinctive properties. For instance, dogkind is characterized by being, amongst other things, carnivorous and mammalkind is characterized by being, amongst other things, warm-blooded. Carnivorousness, thus, is a property of dogkind and warm-bloodedness

---

[8] I discuss these difficulties more fully in my *A Survey of Metaphysics* (Oxford: Oxford University Press, 2002), pp. 348–50 and pp. 382–4.
[9] For further discussion, see my *A Survey of Metaphysics*, pp. 350–2.
[10] I develop a detailed ontology of kinds in my *Kinds of Being: A Study of Individuation, Identity and the Logic of Sortal Terms* (Oxford: Blackwell, 1989).

is a property of mammalkind. Such kinds therefore qualify as objects by being property-bearers, while also qualifying by meeting the condition of individuality: kinds, it seems, have determinate identity conditions and are countable entities.

In fact, the only way in which we might judge kinds to fall short of objecthood is in respect of their ontological dependency upon their particular instances. Dogkind would not exist but for the fact that particular dogs, such as Fido, exist: and, although it is very arguably true that Fido is essentially a dog and so could not have existed if dogkind had not existed, it is still the case that the manner in which the particular instances of kinds depend on kinds for their existence differs from that in which kinds, quite generally, depend for their existence on their particular instances. In this matter, we do well to allow Aristotle to be once more our guide. He called the species to which individual substances belong 'secondary substances', reserving the term 'primary substance' for the individual substances alone. In analogous fashion, we could call *particular* objects, such as Fido, 'primary objects' and call the *kinds* to which such particular objects belong 'secondary objects'. Certainly, so long as we do not overlook the distinction that obtains between particular objects and kinds in virtue of the fact that the latter are universals, it is more illuminating than not to classify both as 'objects'.

5.6 ARISTOTLE'S FOUR-CATEGORY ONTOLOGY RECONSTITUTED

Those who are closely acquainted with Aristotle's metaphysics, and more especially with his work the *Categories*, will realize—even if they have not read previous chapters of this book—that we are now in the process of resurrecting or reconstituting his four-category ontology of two fundamental types of universals and two fundamental types of particulars. (Of course, Aristotle identified more than just four categories altogether, but I am speaking now of the number of ontologically fundamental categories.) The two types of universals are, first, *kinds*—whose particular instances are particular objects—and, second, *properties and relations*, conceived as universals. The two types of particulars are, first, particular *objects* and, second, particular instances of properties and relations—that is, the entities that metaphysicians variously call *tropes*, *property-instances* or *individual accidents*, and which I prefer to call *modes*. The essential relationships between these four categories of entity may be represented by what in earlier chapters I called 'the Ontological Square', as follows (Fig. 5.1), in which I have used the term 'attribute' to denote properties and relations conceived as universals, in order to distinguish properties and relations in this sense clearly from the sense in which they are conceived as particulars.

The Concept of an Object

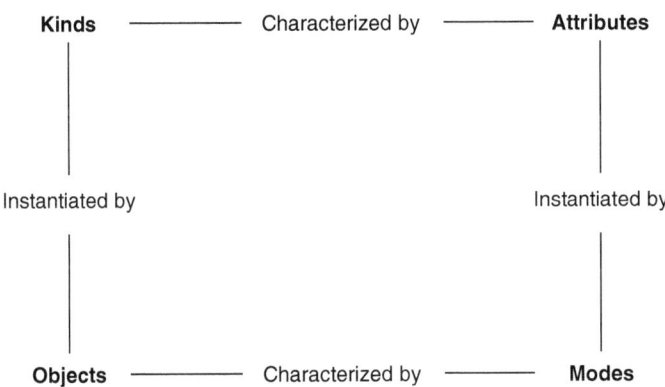

Figure 5.1. The Ontological Square (IV)

It will be noted that in Fig. 5.1 I have restricted the term 'object' to denote only *particular* objects and that, in line with my professed scepticism about the need to include in our ontology 'higher-order' properties, or indeed higher-order entities of any type, I have excluded them from the picture—though this is not to say that the diagram could not be extended to include them if they should after all be wanted.

One feature of the diagram which I should mention again here—although I reserve my fullest discussion of it for Chapter 8 below—is the fact that it allows for attributes to be exemplified by particular objects in two different ways: either in virtue of those objects instantiating kinds which are characterized by the attributes in question, or in virtue of those objects being characterized by modes which instantiate the attributes in question. It will be recalled that, in my view, it is this distinction that ultimately underlies the distinction between *dispositional* and *occurrent* states of objects. For instance, a particular grain of common salt is *water-soluble* because it instantiates a chemical kind, sodium chloride, which is characterized by the property of dissolving in water. Such a grain of salt is *actually dissolving in water*, however, only if it stands to some particular body of water in a relational mode of dissolution. That sodium chloride is characterized by the property of dissolving in water is, in my view, precisely the fact that is reported by the law statement 'Water dissolves sodium chloride'. This makes my view of laws close to that of David Armstrong's, save that where he contends that a law consists in the fact that two or more universals are related by a *higher-order relation* of 'necessitation',[11] I contend that it consists in the fact that a kind is characterized by a certain property or that two or more kinds stand in a certain *first-order* relation—such as the dissolving relation—to one another. These, however, are all matters for more detailed discussion in Part 3 of this book.

---

[11] See D. M. Armstrong, *What is a Law of Nature?* (Cambridge: Cambridge University Press, 1983).

It may be wondered how I can express scepticism about 'higher-order' entities, including higher-order relations, when I speak freely of the ontological relations of instantiation, characterization, and exemplification, all of which may seem to be distinctively 'higher-order' in character. The answer—as I explained in Chapter 3 above—is that these are purely formal relations and so are, in the useful phrase, 'no addition of being'. Even if we accept the existence of relational universals, such as the dissolving relation in which, on my view, water stands to sodium chloride, we should not treat instantiation, characterization, or exemplification as being such relational universals. If we did, we would open up the unwelcome vista of Bradley's regress. For example, when we say that water is characterized by the property of transparency, we should not interpret this as meaning that the kind water and the attribute transparency stand to one another in the relation of characterization, by analogy with the way in which water and sodium chloride stand to one another in the dissolving relation. For this would be to treat the monadic proposition that *water is transparent* as if it were really dyadic, like the proposition that *water dissolves sodium chloride*—and then, of course, to be consistent, we should have to treat the latter as being in reality *triadic*, in which case the proposition that water is transparent would have to be triadic also for the original analogy to hold. Clearly, if we go down this road we shall be driven to deny that any proposition can have a finite adicity, which is patently absurd. As I say, the proper view to take is that ontological relations are purely formal and so are not in any sense elements of being: they are not entities and are not members of any ontological category whatever.

5.7 EVENTS AND PROCESSES

Earlier on, I mentioned that objects, or 'things', are often contrasted by philosophers with *events and processes*, the latter but not the former being said to *happen* or *occur* and to have temporal parts. (Many philosophers do, of course, also hold that persisting objects have temporal parts, but that they do so is at any rate considerably more contentious than that events and processes have temporal parts.) It may be wondered, then, where events and processes fit in my ontological scheme. Here it helps if one asks oneself when it is that an event occurs. I should say that an event occurs when a particular object takes on a particular property, or enters into a particular relation with one or more other particular objects. And by 'particular properties and relations' here I mean, of course, monadic and relational *modes*. For example, an event occurs when a particular object changes from being round to being oblate in shape, as a ball does when it is squashed.

So what *is* this event? What type of entity is it? It is not clear, in fact, that we need to regard it as an entity at all—to *reify* it. Having said what it is for an event of a certain sort to occur, do we need also to say that events *exist*? Perhaps not. But if we do want to identify the event of the ball's becoming oblate with some genuine entity, one possibility would be, quite simply, to identify it with the oblateness mode or trope that the ball acquires. After all, this trope has two key features that we would want to ascribe to the event, assuming that it exists at all. First, like the event, the trope is a datable item. Second, like the event, the trope has a 'subject', namely, the ball. If all we can say with confidence about events is that they occur at times and to objects, then tropes or modes have what it takes to qualify as events. A process, then, might be thought of either as being a temporally extended trope, or as being composed by a temporal succession of different momentary tropes, depending on whether or not the process is a qualitatively unvarying one.

It will be observed that this way of regarding events and processes fits quite comfortably with what I said earlier about the occurrent/dispositional distinction. But the main things that I would wish to emphasize are that I do not believe that we need to think of events and processes as being *objects* of a special type—that is, as being themselves property-bearers—and that I do not believe, either, that events and processes are items belonging to an ontological category of particulars that is at once different from and just as fundamental as the two fundamental categories of particulars that I have already identified, namely, particular objects and their modes.

5.8 NUMBERS AND OTHER ABSTRACT OBJECTS

The next thing I want to do is to return to a distinction that I touched on earlier, to see how it bears on the range of application of the concept of an object: this is the distinction between *concrete* and *abstract* entities. To be precise, there is more than one such distinction to be found in works on metaphysics, but I am at present exclusively concerned with the sense of the term 'abstract entity' in which it is supposed to denote something which does not exist in space or time, paradigm examples being such putative entities as numbers, sets, and propositions.[12] For present purposes I shall simply assume that such entities do exist and are abstract entities in this sense. The question I wish to address is whether or not we should suppose there to be any abstract *particulars* and, more especially, any particular *objects* that are abstract. (Here we must set aside the confusing terminology of some metaphysicians who speak of *tropes* as being 'abstract particulars', meaning thereby something quite other than that they do not exist in space or time.[13]) It might be thought that numbers and sets provide paradigm examples of abstract objects that are particulars, but I have doubts about this. On the one hand, I have some doubts as to whether we should regard sets with full ontological seriousness.

---

[12] See further my 'The Metaphysics of Abstract Objects', *Journal of Philosophy* 92 (1995), pp. 509–24, or my *The Possibility of Metaphysics*, ch. 10.
[13] See Keith Campbell, *Abstract Particulars* (Oxford: Blackwell, 1990).

And on the other hand, while I am relatively happy to acknowledge the reality and abstract nature of numbers, I doubt whether we should regard them as particulars.

Very briefly, my view about numbers—and here I mean the natural numbers—is that they are universals. First of all, although perhaps only to a first approximation, I agree with Locke that number is a property of objects (in opposition to Frege's view, which I shall return to later, that number is a property of *concepts*).[14] Objects, as we have seen, have intrinsic unity—and in virtue of this we may say that each object has the property of *being one*. No single object, of course, has the property of being two, but a plurality of objects can have this property: for example, the moons of Mars, Phobos and Deimos, *are two* in number. Likewise, the planets *are nine* in number, as are the muses. However, I say that this may be only a first approximation to what we should say about number, for the following reason. If being two or being nine really are *properties* of objects, and so belong to my category of attributes, then they ought to have *modes* as their particular instances—and this seems to make little sense. What we can say instead, though, is that, while there are really no such *properties* as being two or being nine, the *numbers* two and nine do really exist and are universals of my other type, that is, that they are *kinds*. One advantage of saying this is that it gives numbers the status of *objects*—albeit not *particular* objects—which is one that they do intuitively have. Numbers, we want to say, are genuinely property-bearers. (Frege, of course, likewise held that numbers are objects, but regarded them as being extensions of concepts—in other words, as being sets and thus as being abstract particulars.)

But if numbers are kinds, what are their particular instances? The answer seems both simple and to leave us with no obvious metaphysical mystery on our hands. We can say that the particular instances of numbers are, in general, *pluralities of objects*—the only exception being the number one, whose particular instances are all the particular objects that there are. Thus, on this view, the Martian moons Phobos and Deimos are, taken together, a particular instance of the number two. But here we should be careful not to be misled by syntax. In saying that Phobos and Deimos are *an* instance of two, we should not be taken to be saying that there is some *one* object which they are and which is an instance of two—for this would be nonsense, implying as it does that there is something that is simultaneously an instance of one and an instance of two. A plurality of objects, such as Phobos and Deimos, is precisely a *many*, not a *one*. The fact that natural language constrains us to use in this context the singular indefinite article and the third person singular form of the verb to be is merely, in Geach's useful phrase, an idiocy of idiom.

Now, however, we face a certain difficulty. This is that the numbers themselves can be numbered. For example, there are four prime numbers less than ten. These

---

[14] See further my 'Identity, Individuality, and Unity'.

four numbers—two, three, five and seven—constitute a plurality which is equinumerous with, say, the plurality constituted by Jupiter's major moons. But whereas those moons are particular objects, I want to say that numbers themselves are kinds and so universals rather than particulars. How, then, can the plurality of prime numbers less than ten be a *particular* instance of the number four? Our difficulty is further compounded by the fact that some pluralities comprise both universals and particulars, such as the foursome consisting of Mars's moons and the first two even numbers.

A simple way out of this apparent difficulty is to resort to talk of sets, regarded as a type of abstract particular whose members can be entities of any type whatever, or at least *objects* of any type whatever (since the axiom of extensionality of set theory requires sets to have their identity determined by the identities of their members, so that those members must themselves possess fully determinate identity conditions). Then we can say that numbers are kinds which have as their particular instances sets of appropriate cardinality, so that, for example, the set {Phobos, Deimos, two, four} would qualify as a particular instance of the number four, even though that set contains the number four itself as a member.[15] If this is the only way to get around the original difficulty, then it provides a good reason to regard sets with full ontological seriousness after all and to accord them the status of abstract particulars. Some would say that we need to do this anyway in order to find a place for the number nought or zero, which most mathematicians simply identify with the so-called empty set. I would only say about this that it is not perfectly clear to me that we need to regard either the number nought or the empty set with full ontological seriousness. However, it may be, in the end, that the proper lesson to draw from the difficulties that we have just been investigating is that the concept of a number is a purely *formal* concept and as such should not be taken to denote any sort of element of being, whether particular or universal, object or property. Here we would again be following a precedent set by Aristotle, who famously insisted that neither Being nor The One is a universal genus.

5.9 FREGE ON OBJECTS AND CONCEPTS

I mentioned earlier Frege's view that number is a property of *concepts* and that the individual numbers themselves are abstract particulars, namely, the extensions of certain concepts (that is to say, sets).[16] His idea is that when we say, for example, that there are exactly *two* Martian moons, we are saying that the concept *moon of Mars* is instantiated exactly twice—in other words, that there is something x and something y, such that x is distinct from y, x and y are both moons of Mars, and

---

[15] This is the view of numbers that I defend in my *The Possibility of Metaphysics*, pp. 223 ff.
[16] See Gottlob Frege, *The Foundations of Arithmetic*, trans. J. L. Austin (Oxford: Blackwell, 1953).

anything *z* such that *z* is a moon of Mars is identical either with *x* or with *y*. We need not go into his view concerning the identity of the individual numbers themselves, because it notoriously fell foul of Russell's paradox. Nor shall I voice any detailed criticisms of his view that numerical adjectives—like the adjective 'two' in the sentence 'There are exactly two Martian moons'—express properties of concepts, since I have done so quite fully elsewhere.[17] My own preferred view, as I have already indicated, is that when we say 'The Martian moons are two in number', what we are saying is that a certain plurality of objects *is a two*, in the sense that that plurality is an instance of the number two, conceived as a kind (a kind of plurality) and thus as a type of universal. But what does call for some further discussion is Frege's own distinction between what *he* calls 'objects' and 'concepts', because it may be far from clear how it relates to the broader concerns of the present chapter.[18]

From my point of view, Frege's way of drawing the object/concept distinction is vitiated by the fact that it rests on a distinction of logical syntax—and I have already emphasized that I regard syntax as a poor guide to ontology. For Frege, objects are the references of singular terms, at least in cases where those terms do have references—and so, by his criteria, at least in those cases in which sentences containing those terms have a truth value. Fregean concepts, on the other hand, are supposed to be the extralinguistic counterparts of predicative expressions, these being what are left of sentences when one or more singular terms are removed from them. By these standards, 'Phobos' denotes an object and '... is a Martian moon' denotes a concept. Notoriously, this way of introducing the object/concept distinction gives rise to the so-called problem of the concept *horse*. The predicate '... is a horse' is supposed to denote a concept—the concept *horse*, or, more idiomatically expressed, the concept of a horse. But the expression 'the concept *horse*' is itself a singular term and so apt only to denote an *object* rather than a *concept*. Frege tries to wriggle out of this difficulty in what is, to my mind, a wholly unsatisfactory fashion.[19]

This problem is, in my view, entirely an artefact of Frege's syntactically driven explication of the object/concept distinction. We should recognize that some singular terms denote objects and some do not, just as some predicative expressions denote properties and some do not. If the predicative expression '... is a horse' denotes anything, what does it denote? Well, if there is such a property as the property of being a horse, then that is what it denotes—and then this, of course, is what is also denoted by the singular term 'the property of being a horse'. But in point of fact I don't believe that there is any such property, because I hold instead that the sortal term 'horse' denotes not a property but a *kind* and that what is asserted when one says that a particular animal *is a horse* is that that animal is an

---

[17] See again my 'Identity, Individuality, and Unity'.
[18] See Gottlob Frege, 'On Concept and Object', *Translations from the Philosophical Writings of Gottlob Frege*, 2nd edn, ed. and trans. P. T. Geach and M. Black (Oxford: Blackwell, 1960).
[19] I say more about this in my *The Possibility of Metaphysics*, pp. 39 ff.

instance of the kind *horse*, or horsekind. Recall here that, on my view, instantiating a kind is very different from possessing (or being characterized by) a property. Frege's account of the object/concept distinction, in addition to the difficulties that I have already mentioned, labours under the difficulty that it fails entirely to register the distinction between properties and kinds. Fregean 'concepts' are somehow meant to embrace both.

5.10 THE ONTOLOGICAL STATUS OF CONCEPTS

But enough of what *Frege* meant by 'concepts'. What should *we* mean by the term? First, we should distinguish clearly between concepts and *universals*, both properties and kinds. The latter are, in general, purely extralinguistic and extramental entities, but *concepts* are not. Concepts, in fact, properly understood, are more like Fregean *senses*. They are ways of thinking of, or intellectually 'grasping', entities. We may, of course, have more than one way of thinking of the same entity and thus different concepts of it. These different concepts may very often be expressed linguistically in different ways. Thus, for example, the adjectives 'triangular' and 'trilateral' very plausibly express or convey different concepts, but these different concepts are plausibly just different ways of thinking of, or grasping, the same extramental geometrical property. Hence, the singular terms 'the property of being triangular' and 'the property of being trilateral' very arguably denote one and the same property, even though they convey different concepts.[20] As for the ontological status of concepts themselves, if they are, as I have just been suggesting, *ways of thinking of entities*, then it would seem that they are mental properties—for properties, quite generally, are appropriately thought of as being *ways* entities are, whether we are talking of properties as universals or properties as particulars (as I shall argue more fully in Chapter 6 below). For example, redness is a way objects can be coloured and squareness is a way they can be shaped. A concept, then, is a way someone can be thinking of an entity. Understood as universals, concepts are mental attributes and understood as particulars they are mental modes. The objects that possess them are thinking subjects, that is, *persons*.

The *concept of an object*, therefore—which has been the topic of this chapter—is the most general way of thinking of any entity that meets the conditions of objecthood that I have proposed (assuming that my proposals are correct). As such, it involves thinking of those entities as being ones which meet those conditions: that is, as being individual property-bearers of order zero. One reason why such a concept is a distinctively ontological one is that it is a purely *formal* concept—unlike, for example, the concept of triangularity. In another terminology, it is a perfectly 'topic-neutral' concept. More importantly, though, the concept of an

---

[20] See further my 'Abstraction, Properties, and Immanent Realism'.

object is not the concept of any *kind* of object, in the way that the concept of a horse, say, is: 'object' does not denote the 'highest' kind of which any particular object is an instance. In Aristotelian terms, *object* is not a universal genus. The concept of an object, then, is not a way of thinking of any distinctive entity or element of being which exists over and above each and every particular object, for there is none. Rather, it is, as I said a moment ago, the most general way there is of thinking of any entity whatever that meets the conditions of objecthood.

6

Properties, Modes, and Universals

6.1 PROPERTIES AND PREDICATES

What are properties? Do any exist? These are surprisingly difficult questions to answer satisfactorily. One might suppose that a property is whatever is denoted by a meaningful predicate and that, since there are many such predicates, there are many properties. But we know that matters cannot be as simple as that, because the supposition that every meaningful predicate denotes a property apparently leads to paradox. Assuming that properties, if they exist, are items that are predicable of other items, it seems that we must suppose the predicate 'is not predicable of itself'—or 'is non-self-predicable', for short—to be meaningful. But if this predicate denotes a property, it presumably denotes the property of being non-self-predicable. And hence, since 'is non-self-predicable' is a meaningful predicate, it must either be true or else be false that the property of being non-self-predicable is non-self-predicable. However, if it is true, then it turns out that that property is, after all, predicable of itself; and if it is false, then it turns out that that property is, after all, not predicable of itself. So we apparently have a contradiction on our hands. There may be ways to evade this result, but at least it shows us that we cannot too lightly assume that every meaningful predicate denotes a property.[1]

None the less, it must be conceded that there is an intimate connection between predicates and properties, not least because our canonical ways of referring to properties—or, perhaps I should more cautiously say, our canonical ways of *attempting* to refer to properties—make use of predicates. The most obvious way of turning a predicate into a singular term which may purportedly be used to refer to a property is to take a predicate, say 'is F', delete the copula, and prefix to what remains the words 'the property of being', to give the noun phrase 'the property of being F'—for example, 'the property of being red', 'the property

---

[1] What we have here, of course, is a version of Russell's paradox. I discuss it further in my 'Abstraction, Properties, and Immanent Realism', in Tom Rockmore (ed.), *The Proceedings of the Twentieth World Congress of Philosophy, Volume 2: Metaphysics* (Bowling Green, OH: Philosophy Documentation Center, 1999), pp. 195–205. For an interesting recent diagnosis and treatment of the paradox, see D. W. Mertz, *Moderate Realism and its Logic* (New Haven: Yale University Press, 1996), pp. 222–5.

of being two miles from the centre of London' and, indeed, 'the property of being non-self-predicable'. Another way, which can only be used conveniently with simple predicates, is again to delete the copula and add to what remains the suffix 'ness', to give the abstract noun '*F*ness'—for example, 'redness', 'roundness', 'tallness' and so forth.[2] But, as we have seen, we have no guarantee that expressions of either form denote anything whatever, even if the corresponding predicates are perfectly meaningful. If we are to answer the questions posed at the outset of this chapter—'What are properties?' and 'Do any exist?'—we need at the very least to provide acceptable accounts of both the *existence conditions* and the *identity conditions* of properties. That is to say, we need to be able to explain satisfactorily what it is, quite generally, for there to *be* such a property as the property of being *F*, or *F*ness. And we need to be able to explain satisfactorily what it is, quite generally, for the property of being *F*, or *F*ness, to be identical with the property of being *G*, or *G*ness—on the assumption that these properties do indeed exist. But it is far from easy to meet either demand.[3]

6.2 UNIVERSALS AND PARTICULARS

It may reasonably be suggested, indeed, that we cannot hope to meet either demand unless we can first determine to what general ontological category properties should be assigned, if indeed they exist at all. Should we conceive of properties as being *universals* or as being *particulars*? Or could it be that the term 'property' is ambiguous and that in one sense it applies to universals of a certain type while in another sense it applies to particulars of a certain type? In any case, what exactly should we understand by the distinction between 'universals' and 'particulars', which are quite as much philosophical terms of art as is the term 'property'? Let me here lay my own cards on the table—or, at least, a few of those cards. I do not believe that the distinction between universals and particulars can be satisfactorily accounted for in spatiotemporal terms. I do not believe, for instance, that a particular may be defined as something that cannot be 'wholly present' in two different places at the same time or, correlatively, that a universal may be defined as something that *can* be 'wholly present' in two different places at

---

[2] Jerrold Levinson has argued that these two types of expression, '[the property of] being *F*' and '*F*ness', denote attributes of quite different sorts, which he calls *properties* and *qualities* respectively: see his 'The Particularisation of Attributes', *Australasian Journal of Philosophy* 58 (1980), pp. 102–15. He remarks that 'The most important characteristic mark of a quality as opposed to a property is variable quantifiability—that is, admitting of *some, more than, less than*' (p. 106). However, he acknowledges that 'it is open for someone to resist seeing quality talk as invoking a notion of variable quantifiable abstract stuff' and that such a person 'will probably seek to replace "A has more φ-ness than B" by the more normal "A has φ in higher degree than B"' (p. 106n). This is a strategy that I am inclined to favour myself.

[3] I propose some ways of meeting these demands in my 'Abstraction, Properties, and Immanent Realism', but I do not rely on these proposals in the present chapter.

the same time. It is not just that I have doubts as to what exactly could be meant by such talk of something's being, or not being, 'wholly present' in two different places at the same time—though I do have such doubts. Rather, it seems to me that the proposal is flawed inasmuch as it rules out by definition the possibility of there being universals or particulars which do not exist 'in' space and time at all— in short, it rules out by definition the possibility of there being *abstract* entities belonging to these categories, in one fairly familiar sense of the expression 'abstract'.[4] It is in this sense that mathematical objects, such as numbers, sets and functions are often characterized as being 'abstract', as opposed to 'concrete', entities. There may, of course, be good reasons to doubt the actual existence of abstract entities in this sense: but so long as such entities *could* exist, and would be either universals or particulars if they did so, it cannot be satisfactory to *define* the distinction between universals and particulars in spatiotemporal terms.

My own view, which is by no means unique to me, is that the distinction between universals and particulars is most satisfactorily captured by appeal to the concept of *instantiation*—a concept which is needed in any case if we are to admit the distinction in question. Particulars instantiate—are, quite literally, *instances of*—universals, at least on the assumption that universals and particulars both exist. Universals, however, may also be thought to instantiate universals, namely, so-called 'higher-order' universals. But the difference between particulars and universals is that, simply in virtue of its being a particular, nothing whatever can instantiate a particular (unless, perhaps, we are prepared to say that every particular trivially instantiates *itself*, but no other particular).[5] This is not, so far, to deny that there may be universals which, as a matter of fact, are not instantiated by anything—'first-order' universals, for example, which have no particular instances. For the proposal is only that every universal, but no particular, is *instantiable*—that is, *can* or *could* have instances. This, it is true, rules out the existence of certain universals that some philosophers might want to include in their ontologies: for instance, on the assumption that there is such a property as the property of being both round and square and that this property is a universal, this is a universal which *could* not have any instances, given that any such instance would have to be both round and square. My proposed way of capturing the distinction between universals and particulars cannot accommodate the existence of such a universal, since it would qualify as a particular according to that proposal. However, this, it seems to me, is a much smaller price to pay than that paid by the spatiotemporal proposal rejected earlier, for that proposal could not accommodate the existence of a whole class of entities—abstract entities—that

---

[4] I discuss this and other senses of the expression 'abstract' in my 'The Metaphysics of Abstract Objects', *The Journal of Philosophy* 92 (1995), pp. 509–24 and in my *The Possibility of Metaphysics: Substance, Identity, and Time* (Oxford: Clarendon Press, 1998), ch. 10.

[5] See further my *Kinds of Being: A Study of Individuation, Identity and the Logic of Sortal Terms* (Oxford: Blackwell, 1989), ch. 3 and J. J. E. Gracia, *Individuality: An Essay on the Foundations of Metaphysics* (Albany, NY: State University of New York Press, 1988).

a great many philosophers are eager to include in their ontologies. Of course, in saying that the spatiotemporal proposal could not accommodate the existence of abstract entities, I am presupposing something that might perhaps be queried, namely, that the distinction between universals and particulars is both exhaustive and exclusive—that everything is either a particular or a universal, but not both. But I hope that that, too, is not unduly controversial. In any case, I shall address any residual worries about this assumption in Chapter 7 below.

6.3 WAYS OF BEING

With the distinction between universals and particulars in place, we can again ask whether properties should be conceived as being particulars or as being universals—or, perhaps, in different senses of the term 'property', as being *both*. In this connection, it is helpful, I think, to draw upon the often-made suggestion that properties may best be thought of as *ways of being*.[6] This is implicitly to deny that properties are *objects* or *things*, at least in a robust and fairly narrow sense of 'object' or 'thing'—the sense which corresponds, more or less, to the traditional concept of 'substance'. The thought, then, is that properties are *ways things are*. That being so, however, it is natural to try to distinguish between a 'way' two or more *different* things may be and a 'way' just one thing is—a 'way' that is necessarily unique to just one thing. And this would correspond, it seems, to the distinction between properties conceived as universals and properties conceived as particulars. According to this suggestion, the property of being red, for instance—assuming there to be such a property—is, when conceived as a universal, a way in which two or more different things may be coloured, such that, each of them being so coloured may be said to be coloured in the *same* way. And by 'in the same way' here is meant, quite literally, 'in the numerically identical way'. At the same time, however, one might want to speak, again quite literally, of the *particular* way in which one thing is coloured and refer to this as, for example, that thing's *particular* redness, or its *particular* property of being red, with the implication that no other thing could be coloured in that very same (numerically identical) way.[7] To put some more of my own cards on the table, I should declare at this point that I believe in the existence of properties, conceived as 'ways', in both of the foregoing

---

[6] This is Jerrold Levinson's suggestion in his 'Properties and Related Entities', *Philosophy and Phenomenological Research* 39 (1978), pp. 1–22; it has been widely followed.

[7] Jerrold Levinson is deeply sceptical about the notion of 'particularized ways', as he explains in his 'Why There Are No Tropes' (forthcoming). He suspects that those who think that there is a sense in which *a*'s being *F* is a numerically distinct property from *b*'s being *F* (where *a* and *b* are themselves distinct objects) have 'shifted attention from properties to states of affairs'. I can only speak for myself in saying that I conceive of *a*'s *particular* property of being *F* as being an entity which, while it depends for its identity upon *a*, does not include *a* as a constituent, in the way that the state of affairs of *a* being *F* is naturally thought to do.

senses—that is to say, I believe in the existence of both *universal* 'ways things are' and *particular* 'ways things are'. The former I simply call *properties*—thus reserving the term 'property' henceforth for a certain type of universal—and the latter I call *modes*, partly out of deference to a long historical tradition which is exemplified, for instance, in the writings of John Locke.[8]

At this stage, something should perhaps be said about *relations*, for I don't want to imply that there are only 'monadic' ways things can be. The property of being red—assuming it to exist—is, inasmuch as it is a universal, a way many things can be, in that each of many things can be that same way. But, in another perfectly clear sense, it is not a way *two or more* things can be, because it is not relational in character. By contrast, the relation of being taller than most certainly is a way two things can be, one with respect to the other. And the relation of being between is a way three things can be, one with respect to the other two. It is natural, then, to think of properties as being, as it were, 'monadic' relations—relations with only one relatum.[9] Of course, in saying this, we must be careful not to confuse the sense in which a property is a 'relation with only one relatum' with the sense in which the relation of identity—assuming there to be such a relation—is a 'relation with only one relatum', in virtue of the fact that identity is a relation in which a thing can stand only to *itself*. I shall not have much more to say in this chapter about polyadic relations, but this is not because I do not think that they are important, much less because I do not think that they exist—for I most certainly think they do. Consequently, I also believe in the existence of the corresponding particulars—relational modes, as they might be called. However, 'monadic' relations or properties and the corresponding particulars—which I shall continue to call simply 'modes'—will do quite enough to keep us occupied for the duration of this chapter.

6.4 INSTANTIATION VERSUS CHARACTERIZATION

I have already mentioned the concept of *instantiation*, in terms of which, indeed, I define the distinction between universals and particulars. Particulars are instances of universals. But it is, in my view, vitally important not to confuse instantiation with the relationship—I won't say 'relation', for reasons to which I shall return—between things and their properties. The latter relationship I like to call 'characterization'. Some philosophers use the term 'exemplification' in this context, but I have another use for that which, as I hope to make plain, requires it to have a different sense from that of the term 'characterization', as I use the latter. Characterization, as I use the term, is a relationship between a particular thing and

---

[8] I say more about Locke's view in my 'Locke, Martin and Substance', *The Philosophical Quarterly* 50 (2000), pp. 499–514.   [9] Compare Mertz, *Moderate Realism and its Logic*, p. 25.

its *particular* properties, or modes. Suppose, for example, that a certain particular flower is red: then—assuming that there is such a property as the property of being red—I want to say that this flower has a particular redness, which is necessarily unique to that flower and which, in my sense of the term, 'characterizes' that flower. (In point of fact, I want to say this only in case the flower is, as I put it, *occurrently* red: but this is a complication that can be ignored for the time being.) The flower's particular redness is a mode—a particular way that flower is—and one which for that reason may be said to 'characterize' the flower. Quite literally, the mode is a particular 'characteristic' of the flower. As I have already indicated, I am reluctant to say that characterization is a *relation* between a particular thing and its modes. For then, it seems, we should have to conceive of a thing and one of its modes as being the relata of a further *relational* mode, which would in turn 'characterize' (in my sense of the term) those two relata, one with respect to the other. And it is easy to see that in this way an infinite regress would be generated which would be at least unwelcome if not fatal.[10] Here, however, we can draw comfort from our earlier observation that not every meaningful predicate need be supposed to denote a property—or, in this case, a relation. Just because 'is characterized by' is a meaningful relational predicate, as it appears in the sentence 'This flower is characterized by its own particular redness', we need not conclude that that predicate denotes a relation in which the flower and its particular redness stand to one another.

The flower's particular redness—a certain mode of the flower which 'characterizes' it—is an instance of the property redness, a universal. In short, modes instantiate properties. But the *flower* whose mode instantiates the property redness does not *itself* instantiate that property. The flower is not literally an *instance* of the property redness—only the mode is that. None the less, inasmuch as the flower is characterized by a mode which instantiates the property, there is a relationship—again, I won't say a *relation*—between the flower and the property and we need to give this relationship a name. This is what I call *exemplification*. However, as will be evident, there are two different ways, or senses, in which something like a flower may be said to exemplify a property such as redness—remembering here that by a 'property' now I mean a certain type of universal.

If the flower is not an instance of the property redness, is it an instance of any universal whatever? Most certainly it is, in my opinion. For example, if the flower is a rose, then it is an instance of the kind *rose*. Kinds are universals, but are not *properties*, where the latter are understood, as I have proposed, as being 'ways things are'. Being red is a way a flower may be, as is being tall or being delicate. But being *a rose* is not a way a flower may be: it is *what* certain flowers are, in the sense that they are particular instances of that kind of thing. To distinguish kinds, thus understood, from properties, I call the former 'substantial universals'. In my view,

---

[10] This is often spoken of as 'Bradley's regress': for an interesting discussion of it, see Mertz, *Moderate Realism and its Logic*, pp. 49–51.

a particular thing's being an instance of a certain substantial universal can never be 'reduced to', or 'analysed in terms of', that thing's being characterized by modes of certain properties.[11] It may indeed be the case that things of certain kinds necessarily exemplify certain properties, but that does not imply that their being of those kinds simply consists in their exemplifying certain properties.

6.5 THE FOUR-CATEGORY ONTOLOGY REVISITED

If the foregoing suggestions are correct, we need to acknowledge the existence of two quite distinct categories of particulars and two quite distinct categories of universals: a fourfold system of fundamental categories which forms the basis of what I call in this book 'the four-category ontology'. It is a system which, according to many commentators, is at least hinted at very early in Aristotle's *Categories*, but which he may or may not have abandoned in subsequent writings.[12] The categories in question are: (1) particular objects or substantial particulars, (2) particular properties or modes, (3) kinds or substantial universals, and (4) properties or non-substantial universals. The system may once again be represented, as in previous chapters, by means of the diagram in Fig. 6.1.

There are some features of Fig. 6.1 which call for further elucidation. First of all, it will be noticed that it represents substantial universals or kinds as being *characterized* by non-substantial universals, that is, by properties, in a way which mirrors the characterization of objects (substantial particulars) by modes. And it

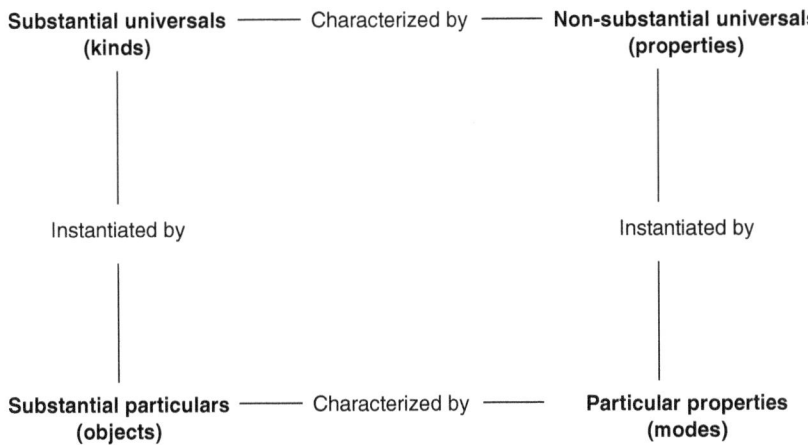

Figure 6.1. The Ontological Square (V)

[11] See further my *Kinds of Being*, pp. 157–8.
[12] For discussion, see Mertz, *Moderate Realism and its Logic*, pp. 98–104.

may be wondered what motivates this idea. The answer is simple. There is a class of statements which express what are sometimes called *generic* propositions and which, taken at face value, seem to be used precisely to say that some kind is characterized by a certain property.[13] An important sub-class of these statements constitute so-called *nomological* statements, or statements of natural law. An example would be the statement—which is false as it happens—'Roses are red', where this is not taken to be equivalent to the universally quantified statement 'All roses are red', nor to the existentially quantified statement 'Some roses are red'. What it appears to be saying—and what, according to the foregoing diagram it *is* saying— is that the property of being red, or redness, characterizes a certain kind of flower, roses. Another, if slightly more stilted, way of saying the same thing would be to say 'The rose is red', where by 'the rose' is not meant any *particular* rose but, once again, the *kind* of which all particular roses are instances. Notice here that the reason why 'Roses are red' is false is not just that there are some particular roses that are not red, but rather that not every *kind* of rose is red. Contrast this with 'Violets are blue', which I take to be true, despite the fact that there may be certain defective or abnormal specimens of the kind *violet* which are not blue. Generic statements of this sort have, in a certain sense, a *normative* character: the most that they tell us about particular instances of a given kind is what *normal* instances of that kind are like, in a certain respect (for example, in respect of their colour).[14]

I concede that it may seem odd, to those who have not reflected on the matter before, to say that a property may characterize a substantial universal or kind: but this is precisely what the syntax of generic statements appears to imply and, in the absence of any compelling reason not to take this appearance at face value, I propose to accept the implication. This proposal has many of the advantages associated with the view that statements of natural law express relations amongst universals without some of the disadvantages of that view.[15] It has the advantage, thus, of making a logical distinction between a statement of law and the corresponding universal generalization—between, for instance, 'Violets are blue' and 'All (particular) violets are blue'—but it does not appeal to the somewhat dubious notion that what 'connects' the universals involved in a law is a 'second-order' relational universal of 'necessitation'. According to the current proposal, what 'connects' the universals involved in a law is simply the familiar relationship—again, I won't say *relation*—of characterization, which equally 'connects' objects to their modes. The welcome implication is that the copula 'is', as it appears in the generic sentence 'The rose is red', has a sense which is intimately related to the sense that it has in the sentence 'This rose is red', where 'this rose' denotes a *particular* rose.

---

[13] See further my 'Noun Phrases, Quantifiers, and Generic Names', *The Philosophical Quarterly* 41 (1991), pp. 287–300.

[14] See further my *Kinds of Being*, ch. 8, and also Chapter 9 below. Compare Alice Drewery, 'Laws, Regularities and Exceptions', *Ratio* 13 (2000), pp. 1–12.

[15] For this alternative view, see D. M. Armstrong, *What is a Law of Nature?* (Cambridge: Cambridge University Press, 1983). For further comparisons and criticisms, see Chapter 9 below.

6.6 EXEMPLIFICATION AND THE COPULA

Here it may be wondered why we shouldn't go further and say that the copula has precisely the *same* sense in those two sentences. The answer has to do with the notion of exemplification, which I introduced earlier, and can be readily understood by looking again at Fig. 6.1. I remarked earlier that a substantial particular, or 'object' (as I am now using that term), never *instantiates* a non-substantial universal, or property—though it may be *characterized* by a mode of such a universal, in which case it may be said to *exemplify* the universal. However, from Fig. 6.1 we can see that there are clearly *two* different types of relationship between objects and properties, only one of which we have identified so far in this chapter. This first type of relationship obtains when an object is characterized by a *mode* of a certain property. The second type of relationship obtains when an object instantiates a *kind* which is characterized by a certain property. Thus, there are two different senses in which an object may be said to exemplify a certain property, or non-substantial universal: two different senses, thus, in which a particular flower may be said to 'be red' and so two different senses of the sentence 'This flower is red'.

This, at least, is what our diagram informs us. But does the proposal correspond to any ambiguity which can actually be discerned in such a sentence of natural language? I think it clearly does. For the predicate 'is red' in such a sentence can be understood either in an *occurrent* or in a *dispositional* sense.[16] If, as is the case in some languages other than English, we used verbs to ascribe colours to objects instead of adjectives, the distinction might be more obvious and would be registered at the level of surface syntax: we would distinguish, thus, between the sentences 'This flower *reds*' (dispositional) and 'This flower *is redding*' (occurrent), just as we distinguish between 'This sugar cube *dissolves* in water' (dispositional) and 'This sugar cube *is dissolving* in water' (occurrent). According to our diagram, the root of the occurrent/dispositional distinction is quite straightforward. For an object to be *occurrently* F is for that object to be characterized by some mode of Fness, whereas for an object to be *dispositionally* F is for that object to instantiate some kind that is characterized by Fness. In view of our earlier remarks, this proposal connects dispositions and laws in an obvious way: for an object to possess a certain disposition is just for that object to be an instance of a kind that is involved in a corresponding law.[17] Thus, the sugar cube possesses the disposition to dissolve in water because it is an instance of a kind—the *sugar* kind—of which it is a characteristic that it dissolves in water: as we say, 'Sugar dissolves in water', meaning thereby to express a law concerning the kind *sugar*.

This, then, is why I implied earlier that the copula in the two sentences 'The rose is red' and 'This rose is red' could not be taken to have exactly the same sense

[16] See further my *Kinds of Being*, p. 140.   [17] Compare my *Kinds of Being*, p. 170.

in each. In the latter sentence, the predicate 'is red' has both an occurrent and a dispositional reading, whereas there is only one natural reading of the former sentence, in which it purports to express a law. If pressed to say whether this reading is 'dispositional' or 'occurrent', we should say that it is 'dispositional'. And, indeed, there is much to be said for the idea—which is by no means new—that lawlike statements tell us how objects of certain kinds are normally *disposed* to behave or appear.[18] Think—to use an example that I have cited before—of Kepler's laws of planetary motion, one of which states that planets orbit the sun in ellipses with the sun at one focus. Clearly, even if relativistic physics now implies that this law is not strictly true, its truth was never impugned by the fact that, for example, a planet might actually be deflected from its orbit by a passing comet. This is because the law did not purport to describe how each particular planet is actually moving, but only how planets are disposed to move under the influence of the sun's gravitational field (though Kepler himself, of course, would not have put it quite that way).

6.7 MODES VERSUS TROPES AND TRANSCENDENT UNIVERSALS

So far, we have seen that the four-category ontology, as represented by our earlier diagram, seems to accommodate many features of our everyday and scientific talk of objects and their properties. None the less, in the eyes of philosophers with a penchant for desert landscapes, it will seem to be an extravagant ontology. Can it be satisfactorily defended against such critics? I believe so. At the risk of repeating one or two things said elsewhere in this book, let me now cite some crucial points of difference between the four-category ontology and certain of its rivals and try to explain why I think that the four-category ontology has the edge over them.

Some philosophers who are happy to include universals in their ontology will think it extravagant of me to include what I call modes as well. Indeed, many philosophers who currently debate over the ontological status of properties assume that the only choice before us—given that we at least accept that properties do exist—is between conceiving properties as universals and conceiving them as particulars, with self-styled 'trope' theorists taking the latter view.[19] In this connection, I should remark that one reason why I prefer the term 'mode' to the term 'trope' to denote something in the category of *particular* properties, is that this preserves the traditional association between such entities and the correlative category of particular *substances*, or 'objects'. Trope theorists typically maintain

---

[18] See again my *Kinds of Being*, ch. 8, and also my 'What *is* the "Problem of Induction"?', *Philosophy* 62 (1987), pp. 325–40.

[19] For a defence of trope thory, see Keith Campbell, *Abstract Particulars* (Oxford: Blackwell, 1990).

that what we are apt to call 'objects' are no more, in reality, than 'bundles' of 'compresent' tropes. In contrast, I contend that modes are 'particular ways objects are', and as such are ontologically dependent upon objects in a much stronger sense than, according to a trope theorist, any trope can be ontologically dependent upon other tropes in a bundle of compresent tropes. Contrary to the supposition of some trope theorists, this does not commit me to some untenable doctrine of 'featureless substrata' or 'bare particulars' in order to explain in what sense an object, or particular substance, is 'more' than just a collection of particular properties.[20] According to my conception of objects, an object is not a complex which is somehow constituted by a collection of particular properties *together with* some further entity which is itself neither a particular property nor a propertied object. The mistake is to suppose that an object is even *partially* constituted by its particular properties, as this inverts the true direction of ontological dependency between object and property. Particular properties are no more (and no less) than *features* or *aspects* of particular objects, which may indeed be selectively attended to through a mental process of abstraction when we perceive or think of particular objects, but which have no being independently of those objects and which consequently cannot in any sense be regarded as 'constituents' of objects. In this respect, the particular properties of an object differ radically from its *parts*, if it has any, for these are just further objects with particular properties of their own.[21]

Setting aside trope theory, then, for the foregoing reasons, I have to contend with those philosophers who accept that particular objects are not reducible to bundles of compresent tropes, but who none the less maintain that such objects only possess properties conceived as universals. For these philosophers, there are no such entities as 'particular properties', or, as I put it earlier, 'particular ways objects are'. Properties may still be regarded as 'ways object are', but only as (identical) ways *many* objects may, at least in principle, be. It may be conceded that there may be properties which are exemplified by only one object, such as, perhaps, the property of being the first man to set foot on the moon (if we seriously think that there is such a property). But, even so, these properties are taken not to be dependent for their very identity upon the objects whose properties they are, in the way that particular properties or modes are naturally taken to be. Thus, even though, of necessity, there is only one object that can exemplify the property—conceived as a universal—of being the first man to set foot on the moon, that very property could have been exemplified by a different object from the object, if any, that actually exemplifies it. By contrast, this flower's particular redness—assuming there to be such a thing—could not have been possessed by any other object whatever, because its being *this* flower's redness is at least partly what makes it the particular property that it is.

---

[20] See further my 'Locke, Martin and Substance'.
[21] Compare C. B. Martin, 'Substance Substantiated', *Australasian Journal of Philosophy* 58 (1980), pp. 3–10.

But the challenge, now, is to say why we should believe in the existence of such entities as this flower's particular redness *in addition* to the corresponding universals, such as the property of being red. One reason that may be offered is that universals require particular instances—that there cannot be *uninstantiated* universals—and yet that particular objects, such as this flower, are not *instances* of properties, conceived as universals: rather, objects 'exemplify' such universals. Perhaps, however, this appeal to the distinction between instantiation and exemplification will be deemed to be question-begging in the present context. Furthermore, some philosophers do, of course, believe in the existence of uninstantiated universals, so that we cannot unquestioningly adopt the opposite point of view. Even so, it may be pointed out that if universals can exist uninstantiated, then they must apparently be *abstract* objects, in the sense invoked earlier in this chapter—that is to say, they must be entities which do not exist 'in' space and time. Universals thus conceived are often described as 'transcendent' universals, or universals *ante rem*, as opposed to 'immanent' universals, or universals *in rebus*. And there is undoubtedly a problem in supposing that concrete objects, which exist 'in' space and time, could have as their properties only transcendent universals. For when we *perceive* an object, we perceive some of its properties—how else, in the end, could we know what properties objects have? And when objects interact causally with one another, the manner of their interaction is determined by their properties. Indeed, since perception itself is at bottom just a species of causal interaction between one object (the perceiver) and another (the object perceived), the first point is embraced by the second. And then the problem is that entities which do not exist 'in' space and time—the objects of mathematics being the paradigm examples—are necessarily causally inert, so that transcendent universals cannot play the role in perception and causation that at least some of the properties of objects are required to play.

6.8 TWO CONCEPTIONS OF IMMANENCE

This line of reasoning may suffice to show that properties conceived as *transcendent* universals are not enough, so that if that is what properties conceived as universals should be taken to be, then we need particular properties, conceived as concrete entities existing 'in' space and time, either in addition to or instead of such universals. But it does not suffice to show that properties conceived as *immanent* universals are not enough. However, the doctrine of immanence is subject to two different interpretations—and, I believe, on one interpretation it borders on incoherence while on the other it leaves us with a conception of universals which is no better, in the respects now crucial, than is the conception of universals as being transcendent. According to what I shall call the 'strong' doctrine of immanence, universals exist 'in' space and time in the sense that they may be quite literally 'wholly present' in many different places at the same time.²² That is to say, the very self-same universal redness which is exemplified both by this flower and by that different flower is, according to this view, located *in its entirety* both in the same place as this flower and in the same place as that flower. It is not, then, that *part* of the universal is located in the one place and (another) *part* of it in the other place, for the universal is not considered to have parts in this sense. However, it is difficult to understand how anything meeting this description could literally be true. For the two flowers are undoubtedly in wholly different locations. And yet we are to suppose that all of the first flower coincides spatially with all of the universal and, at the same time, that all of the universal coincides spatially with all of the second flower. (I am, of course, assuming for the sake of the example—I take it uncontroversially—that both flowers are wholly red in colour.) But then it is hard to see how the first flower could fail to coincide spatially with the second flower. It is no use just insisting that this need not be so, on the grounds that universals behave quite differently from particulars where matters of spatiotemporal location are concerned. For it needs to be explained to us *how* they can behave so differently, despite genuinely being located in space and time. And I have never yet come across a satisfactory explanation of this purported fact. As it stands, then, it seems to be nothing more than a piece of unsupported dogma.

Indeed, we can press the case further against the adherents of this dogma, by challenging them to say how their thesis—that 'all' of a universal may simultaneously be located in two different places—presents a picture of the world that is intelligibly different from that presented by the trope theorist, who says instead that what may be simultaneously located in two different places are two distinct but exactly resembling tropes. How would the world present itself as being any different, to the minutest inspection—even, as it were, to the mind of God— according to the two allegedly different accounts? The alleged distinction appears to be a distinction without a meaningful difference. So the upshot of our inquiries is that, if the 'strong' doctrine of immanence is advanced in opposition to a transcendent conception of universals, then this leaves its adherents either with an inexplicable mystery which borders on incoherence or else with a doctrine which, to the extent that it is intelligible at all, seems to collapse into the trope theorist's conception of properties as being one and all particulars.

There is, however, also a 'weak' doctrine of immanence to be taken into consideration. This just amounts to an insistence upon the instantiation principle—the principle that every existing universal is instantiated.²³ Applied to a universal such as the property of being red, it implies that this universal must have particular instances which exist 'in' space and time, but it doesn't imply that the universal itself must literally exist 'in' space and time. Assuming this doctrine

---

²² For this notion, see, for example, D. M. Armstrong, *Universals: An Opinionated Introduction* (Boulder, CO: Westview Press, 1989), pp. 98–9.
²³ I endorse this principle in my 'Abstraction, Properties, and Immanent Realism'.

to be correct, we must now return to the question of what, exactly, these 'instances' are, in the case of a property such as redness. Suppose it is contended, contrary to the tenets of the four-category ontology, that the particular instances of redness are just *objects*, such as this flower and that flower, rather than what I have been calling *modes* of such objects—this flower's particular redness and that flower's particular redness. Then again we have difficulty, I suggest, in understanding how an object's properties could have any bearing upon its causal transactions with other objects and how, more especially, we could discover an object's properties by perceiving it. For if there is, so to speak, nothing *about* the object— no discriminable feature or aspect of it—which relates it to each property (conceived as a universal) which the object is said to instantiate, so that it is only the object *simpliciter* or 'holus-bolus' which instantiates each of its properties, then it once again becomes a mystery how its being so related to those universals makes any difference *for us* and other denizens of the sublunary world of things in space and time.[24] It is for this reason, then, that I consider that, where properties are concerned, universals are not enough, whether these are conceived as transcendent or as (weakly) immanent. A belief in the existence of properties, conceived as universals, demands a belief in the existence of *modes*, conceived as particular instances of those properties which characterize objects exemplifying the properties in question.

In this chapter, I have gone a little way towards answering our two initial questions—'What are properties?' and 'Do any exist?'—but have also left much unsaid. I have not even attempted to formulate a comprehensive account of the existence conditions and identity conditions of properties, which would need to be done in order to answer our two questions fully. But at least I hope to have provided some good reasons for thinking that this is a task worth pursuing.

---

[24] See also my *The Possibility of Metaphysics*, pp. 204–5. Responding to my claims there, Jerrold Levinson suggests, in 'Why There Are No Tropes', that instead of speaking—as I would wish to do—of seeing a certain leaf's *particular greenness*, it suffices to speak of seeing a certain *state of affairs*, the leaf's being green, or *that* the leaf is green. However, I do not regard seeing *that* as being literally a kind of *seeing*. I consider 'I see that *p*' to express a visually based judgement that *p*, rather than a report that one is seeing something: see my *An Introduction to the Philosophy of Mind* (Cambridge: Cambridge University Press, 2000), p. 132. And then my point would be that such a judgement needs to be based on something distinctive and relevant that *is* seen, quite literally, with the leaf's greenness being the obvious candidate in the present case. Moreover, it must be the leaf's *particular* greenness, because seeing involves a causal relation between the seer and the entity that is seen—and only spatiotemporally located particulars can enter into causal relations. Levinson offers as another candidate the leaf's green *pigment*, but I am not entirely convinced that this chemical substance— chlorophyll—is (ordinarily, at least) something that is *seen* when one sees that the leaf is green: rather, it is what makes the leaf appear green, or gives it its green colour. Moreover, even if the pigment *is* seen, it is at best just another green *thing*, like the leaf itself—and merely seeing something that is green is not enough to provide a basis for the visual judgement that one sees *that* the thing in question is green: seeing something *about* the thing must provide that basis, its (particular) greenness once again being the obvious candidate.

7

Ramsey's Problem and its Solution

7.1 THE CHALLENGE OF RAMSEY'S PROBLEM

Frank Ramsey's justly famous paper 'Universals', first published in *Mind* in 1925, opens with the following words: 'The purpose of this paper is to consider whether there is a fundamental division of objects into two classes, particulars and universals'.[1] At least two things emerge clearly from the paper. One is that, in Ramsey's view, 'the distinction between particular and universal [is] derived from that between subject and predicate'.[2] The second is that he considers the theory of universals to be a 'great muddle', which arises from a failure to heed the distinction between names and so-called *incomplete symbols*.[3] If Ramsey is right, then, metaphysicians who believe in the distinction between universals and particulars are victims of a misunderstanding about how language works. The problem that he poses for such metaphysicians is the problem to which I refer in the title of the present chapter as 'Ramsey's problem'. The problem they face is to justify their belief in the distinction between universals and particulars in the light of the difficulties which Ramsey seems to raise for it. Since I count myself amongst the metaphysicians in question, it is a problem that I too need to confront.

I may begin by repudiating, for my own part, Ramsey's suggestion that the distinction between particular and universal is *derived* from that between subject and predicate. That is to say, I have no inclination whatever to contend that what either motivates or justifies the ontological distinction between particular and universal is this syntactical distinction between subject and predicate. This is partly because, as I shall explain in a moment, I think that Ramsey conflates the universal–particular distinction with another quite different ontological distinction. But it is also because I have no inclination whatever to contend that ontological distinctions can be founded upon syntactical ones or, more generally, that metaphysical distinctions can be founded on linguistic ones. It might be thought that this entitles me to ignore Ramsey's paper altogether as presupposing something that I entirely reject. But that is not so—first, because at least some of

---

[1] See F. P. Ramsey, 'Universals', in his *The Foundations of Mathematics and Other Logical Essays* (London: Kegan Paul, 1931), p. 112.   [2] Ibid., p. 128.
[3] Ibid., p. 134.

the difficulties that Ramsey raises for theories of universals seem not to depend crucially on the rejected presupposition and, second, because it is incumbent upon me to offer an alternative justification for the distinction between universals and particulars, given that I repudiate a syntactically based one.

7.2 RAMSEY'S CONFLATION OF TWO DIFFERENT DISTINCTIONS

Before proceeding further, I want to draw attention to an important feature of Ramsey's discussion which is apt to be overlooked and to which I alluded a moment ago. This is that throughout his paper he seems to conflate two purported ontological distinctions which should, very arguably, be kept clearly apart. One is the distinction between universals and particulars, which is the official subject of his paper. The other is the distinction between properties and relations, on the one hand, and the bearers of properties and relata of relations on the other. Apart from a brief and dismissive reference to G. F. Stout, Ramsey takes no notice of the view that properties and relations may themselves be particulars rather than, or as well as, universals, announcing that he will discuss only 'the more usual opinion to which Mr Russell adheres' that 'the class of universals is the sum of the class of predicates and the class of relations'.[4] Incidentally, by the 'class of predicates' Ramsey here clearly means the class of *properties*. It is a confusing feature of Ramsey's paper that he frequently uses linguistic and ontological terminology interchangeably, such as the terms 'predicate' and 'property'. Ramsey himself was undoubtedly not a victim of confusion on this score, but it is an unfortunate feature of his paper none the less, in view of the fact that one of his main aims is to argue that a purported ontological distinction has been illegitimately derived by metaphysicians from a linguistic one. For by using linguistic and ontological terminology interchangeably he helps to lend this claim more plausibility than, perhaps, his arguments warrant.

Let me return to the point just made, that Ramsey conflates the distinction between universals and particulars with the distinction between properties and relations, on the one hand, and the bearers of properties and relata of relations on the other. To avoid unnecessary prolixity, let us refer to the latter distinction simply as the distinction between properties and property-bearers—although we might more fittingly refer to it as the distinction between relations and relata, bearing in mind a point that Ramsey himself acknowledges, that properties may be thought of as monadic relations.[5] Nominalist metaphysicians—understanding these to be those who deny the existence of universals—hold that everything that exists is particular, but disagree amongst themselves concerning the distinction

[4] Ramsey, 'Universals', p. 113.   [5] Ibid.

between properties and property-bearers. Some assume that properties, if they existed, would have to be universals, and so conclude that properties do not exist. Others maintain that properties do exist, and so are one and all particulars. Those who hold the latter view further divide into two camps. One camp—whose members are nowadays known as *trope theorists*—believe that all particulars are properties and consequently that there is no ontological distinction to be drawn between properties and property-bearers.[6] The other camp upholds the latter distinction. According to the trope theorists, so-called property-bearers are merely 'bundles' of tropes. Their opponents hold that a property-bearer is a particular that is not to be identified with any property or collection of properties: rather, it is something belonging to an altogether distinct and equally fundamental ontological category. Sometimes this sort of particular is referred to as a 'substance', sometimes as a 'substratum' or 'bare particular'—and these different terms reflect different views about the exact nature of property-bearers. A further complication, which will be seen to be of some importance in due course, is that some metaphysicians hold that properties themselves can be bearers of so-called 'higher-order' properties, so that for these philosophers the only property-bearers that are not themselves properties are the bearers of 'first-order' properties.

It will already have emerged from what I have just been saying that the metaphysical landscape is a good deal more complex than might be suggested by a cursory reading of Ramsey's paper. While ostensibly attacking the distinction between universals and particulars quite generally, he is really attacking, in effect, a quite specific ontological position which admits of many rivals. The position in question is one which holds that all properties are universals and all particulars are property-bearers. Setting aside the fact that this position may allow that properties themselves are bearers of so-called higher-order properties, it is a position which encourages the conflation that Ramsey himself indulges in—the conflation of the distinction between universals and particulars with the distinction between properties and property-bearers. But, as we have remarked, there are other positions which do not encourage this conflation.

One such position which I have not so far mentioned—but which I mention now because I favour it myself—is one which acknowledges the existence of no less than *four* fundamental ontological categories: two categories of universals and two categories of particulars. One of these categories of universals is the category of properties and relations, conceived as universals. The particular instances of these universals—otherwise known as tropes—are then taken to constitute the membership of one of the two fundamental categories of particulars. The other fundamental category of particulars is the category of particular property-bearers, otherwise known as 'substances' (although in fact this term, as it is ordinarily understood, is perhaps too narrow in scope for the present purpose). The fourth fundamental category is the remaining category of universals: the category of

---

[6] See, for example, Keith Campbell, *Abstract Particulars* (Oxford: Blackwell, 1990).

universals whose particular instances are particular property-bearers. These might be called (although, again, the term is perhaps too narrow in scope) *substantial universals* or *kinds*.

7.3 RAMSEY'S OBJECTIONS TO THE UNIVERSAL–PARTICULAR DISTINCTION

We now need to look at some of Ramsey's arguments that are designed to call into doubt the distinction between universals and particulars, recalling here his contention that this distinction is 'derived' from the distinction between subject and predicate. His main and best-known objection is that philosophers who derive the first of these distinctions from the second make a questionable assumption, namely, 'They assume a fundamental antithesis between subject and predicate, that if a proposition consists of two terms copulated, these two terms must be functioning in different ways, one as subject, the other as predicate'.[7] He famously takes the example of the proposition 'Socrates is wise', and observes that if 'we turn the proposition round and say "Wisdom is a characteristic of Socrates", then wisdom, formerly the predicate, is now the subject'.[8] He then remarks: 'Now it seems to me as clear as anything can be in philosophy that the two sentences "Socrates is wise" and "Wisdom is a characteristic of Socrates" assert the same fact and express the same proposition'[9]—and goes on to reason as follows:

Now of one of these sentences 'Socrates' is the subject, of the other 'wisdom'; and so which of the two is subject, and which predicate, depends upon what particular sentence we use to express our proposition, and has nothing to do with the logical nature of Socrates or wisdom, but is a matter entirely for grammarians... Hence there is no essential distinction between the subject of a proposition and its predicate, and no fundamental classification of objects can be based upon such a distinction.[10]

The first thing to note about what Ramsey says here is that it exhibits the same laxity in the use of ontological and linguistic terminology that we noted earlier in the case of the terms 'property' and 'predicate'. Thus Ramsey slides between using the term 'proposition' to denote a linguistic item—a sentence—and using it to denote something that one or more sentences may be taken to *express*. Again, at one point he speaks of *wisdom*—an extralinguistic entity—as being a 'subject', while later he speaks of the *word* 'wisdom' as having this status. To avoid all possible confusion on this score, we would do well to use the terms 'subject', 'predicate' and, of course, 'sentence' solely to denote linguistic items and the term 'proposition' solely to denote something extralinguistic—something that one or more sentences may be taken to express or to have as their 'meaning' (though the slipperiness of the latter term makes it dangerous in this context).

[7] Ramsey, 'Universals', p. 116. [8] Ibid. [9] Ibid. [10] Ibid.

The crux of Ramsey's reasoning in the above passage is that the two sentences 'Socrates is wise' and 'Wisdom is a characteristic of Socrates' express the same proposition and 'have the same meaning'. To see why, we should note that Ramsey holds that the identity of a proposition depends on the identity of its 'constituents', as emerges from what he says a little later in the paper. It is worth looking at this later passage more closely. Ramsey is commenting on the view that, given a relational affirmation of the form '*aRb*',

[W]e can distinguish three closely related propositions; one asserts that the relation R holds between the terms a and b, the second asserts the possession by a of the complex property of 'having R to b', while the third asserts that b has the complex property that a has R to it.[11]

Concerning this view, he remarks, by way of presenting a *reductio ad absurdum* of it:

These must be three different propositions because they have different sets of constituents, and yet they are not three propositions, but one proposition, for they all say the same thing, namely that a has R to b. So the theory of complex universals is responsible for an incomprehensible trinity, as senseless as that of theology.[12]

Now, as we have seen, Ramsey considers that we cannot take the sentences 'Socrates is wise' and 'Wisdom is a characteristic of Socrates' to express two different propositions. It seems to him evident that the two sentences have the same meaning. Hence, he believes, they must have the same constituents. And on the assumption that the proposition in question is an atomic one, these constituents could only be, it would seem, Socrates and wisdom—although Ramsey himself, writing here under the influence of Wittgenstein's *Tractatus*, is of the opinion that we are unable to discover genuinely atomic propositions by analysis and so do not know what their constituents are. Suppose, however, that the proposition in question were atomic and that its constituents were Socrates and wisdom. Then notice that there could not be a *different* atomic proposition with these same constituents, as it seems there could in the case of a proposition whose constituents were a, R and b. If R is an asymmetrical relation, we can distinguish between the proposition that a has R to b and the proposition that b has R to a. We cannot do this in the case of a proposition containing only two 'terms'. Consequently, according to Ramsey's way of thinking, 'Socrates is wise', or its equivalent 'Wisdom is a characteristic of Socrates', cannot incorporate in its meaning any kind of asymmetry between Socrates and wisdom, of the sort presupposed by those metaphysicians who hold the former to be a 'particular' and the latter to be a 'universal'.

It might be supposed that there is a way out of this apparent difficulty, by holding that 'Socrates is wise' really expresses a relational proposition or fact, to wit, the fact that Socrates stands in the relation of *being characterized by* to wisdom, so that while 'Wisdom is a characteristic of Socrates' does indeed express

[11] Ibid., p. 118. [12] Ibid.

the same proposition—rather as 'Brutus killed Caesar' expresses the same proposition as 'Caesar was killed by Brutus'—we can none the less discern a genuine asymmetry between Socrates and wisdom. For now, it might be said, we can distinguish between this true proposition and another false proposition with the same constituents, namely, the proposition expressed by the sentence 'Socrates is a characteristic of wisdom'. Interestingly enough, we cannot turn the latter into an equivalent English sentence of the form 'Wise is Socrates', for the latter is simply a stilted, but perfectly acceptable, way of rephrasing 'Socrates is wise'. The problem with this strategy, however, is that it seems to fall foul of Bradley's notorious regress, to which Ramsey himself refers. He cites Russell as approving of Bradley in this connection and maintaining that 'a relation between two terms cannot be a third term which comes between them, for then it would not be a relation at all, and the only genuinely relational element would consist in the connections between this new term and the two original terms'.[13]

Ramsey also cites Russell as holding, instead, that universals differ from particulars in being 'incomplete'—which was, of course, essentially the view of Frege, expressed in terms of his own distinction between 'concepts' and 'objects', with the former described by him as being distinctively 'unsaturated'. But Ramsey then complains that 'In a sense, it might be urged, all objects are incomplete; they cannot occur in facts except in conjunction with other objects' and asks, 'In what way do universals do this more than anything else?'.[14] (By 'objects' here, he does not mean, of course, precisely what Frege means by 'objects'.) Ramsey also takes note of W. E. Johnson's view that in a proposition 'the connectional or structural element is not the relation but the characterizing and coupling ties', but immediately observes that 'these ties remain most mysterious objects'.[15]

The upshot of all this seems to be, at least in Ramsey's view, that there really cannot be anything in the nature of a two-termed atomic proposition that either engenders or reflects any asymmetry between the two terms. If he has sympathy for any of his contemporaries' views, it would seem to be for the view of Wittgenstein, whom Ramsey describes as holding that in an atomic proposition or fact 'neither is there a copula, nor one specially connected constituent, but..., as he expresses it, the objects hang one in another like the links of a chain'[16]— though this metaphor seems no less obscure than others we have just encountered, in talk of 'incomplete' objects or characterizing 'ties'. On the Wittgensteinian view, if 'Socrates is wise' expressed an atomic proposition—which, of course, he would deny that it could—its constituents, Socrates and wisdom, would 'hang one in another like the links of a chain': and, like links in a chain, they would be functionally equivalent, that is, entirely symmetrical with respect to the 'linkage' between them.

---

[13] Ramsey, 'Universals', p. 115.   [14] Ibid.   [15] Ibid.   [16] Ibid., p. 121.

7.4 THE LESSON OF RAMSEY'S ARGUMENTS

It is striking that, being reminded of all of these positions—Russell's, Frege's, Johnson's, and Wittgenstein's—eighty years after the publication of Ramsey's paper, analytical metaphysics may not seem to have made much advance since then: for we still hear appeals to these ideas of 'incompleteness' or 'unsaturatedness', characterizing 'ties' and the like in present-day attempts to account for what is sometimes called 'the unity of the proposition' or, more recently, the special 'non-mereological' mode of composition allegedly peculiar to states of affairs. What, then, should the modern advocate of the universal–particular distinction say in response to Ramsey's objections to it?

First, we should briefly take note of another way in which Russell, as reported by Ramsey, attempted to ground the distinction, namely, by appealing to the fact that properties and relations (which Russell, of course, took to be universals) seem to have a fixed 'adicity', whereas particulars—that is to say, property-bearers (which Russell took all particulars to be)—do not. As Ramsey puts it: '[Russell] says that all atomic propositions are of the forms $R_1(x)$, $R_2(x, y)$, $R_3(x, y, z)$, etc., and so can *define* individuals [i.e., particulars] as terms which can occur in propositions with any number of terms; whereas of course an n-termed relation could only occur in a proposition with $n + 1$ terms'.[17] However, Ramsey complains that this already presupposes the distinction between particulars and universals and, equally illicitly, presupposes that we can know the forms of atomic propositions. It is not clear that either of these objections has much force as it stands and, indeed, in a note written in the following year Ramsey retracted his claim that it is impossible to discover atomic propositions by analysis and conceded that we might in principle be able to discover them to have the forms proposed by Russell—although he still insists that none of this is knowable *a priori*.[18]

But, in any case, other difficulties seem to beset Russell's proposal. One is that properties and relations themselves may perhaps have so-called higher-order properties or stand in higher-order relations, in which case, for example, a property or monadic relation might apparently occur in an atomic proposition containing not just two but, say, three terms. To cite a putative present-day example, Armstrongian laws are supposed to be like this, having (in the simplest case) the form '$N(F, G)$', where 'N' denotes a second-order relation of 'necessitation' whose relata are the first order monadic universals F and G.[19] Another difficulty is that some relations seem to be 'multigrade', that is to say, they seem not to have a

---

[17] Ibid., p. 133.
[18] See F. P. Ramsey, 'Note on the Preceding Paper', in his *The Foundations of Mathematics*, pp. 135–7.
[19] See David M. Armstrong, *What is a Law of Nature?* (Cambridge: Cambridge University Press, 1983).

fixed 'adicity'. A putative example would be the relation of composition in which parts stand to the whole which they collectively compose: for a whole may have any number of parts and it seems absurd to suppose that, for each number of parts that a whole may have, a different relation of composition must be involved. The first of these difficulties might be thought to be overcome by stipulating that 'particulars' are, as it were, 'zero-order' entities—or else, more drastically, by denying the existence of 'higher-order' universals (see Chapter 5 above). However, both suggestions seem to presuppose already the very distinction between universals and particulars which is now at issue and so cannot apparently solve the problem in hand, however meritorious either of them might be in its own right. In any case, I shall not pursue the Russellian proposal any further here, because it seems to face too many potential obstacles.

In my view, the real lesson of Ramsey's paper is that there is not much prospect of justifying the universal–particular distinction within the framework of an ontology which takes 'states of affairs' or 'facts' to be the basic building blocks of reality—where states of affairs are taken to be items isomorphic with propositions, either because they just *are* propositions (facts, on this view, being true propositions) or because they are the truthmakers of propositions.[20] If states of affairs are taken to be the basic building blocks of reality, then their 'constituents' are, for that very reason, *not* taken to be basic. The 'constituents' of a state of affairs cannot, on this view, be taken to be items of an ontologically independent character which somehow 'come together' to compose, in some mysterious way, the state of affairs in question—somewhat in the manner in which bricks compose a wall, by being cemented together in rows. For the bricks exist independently of the wall and are ontologically more basic than it is. Rather, the 'constituents' of states of affairs can only be thought of as being *abstractions from*, or *invariants across*, a totality of states of affairs—the sort of totality that, according to a states of affairs ontology, constitutes a possible *world*. (The actual world, on this sort of view, is—in Wittgenstein's memorable phrase—the totality of facts, not of things.) Hence, it seems, on this sort of view, constituents of states of affairs can only be partitioned into different ontological categories if systematic differences can be discerned between the ways in which different classes of constituents contribute to the structure of the states of affairs containing them. If it could be shown, for instance, that there are two classes of constituents, U and P, such that members of U only recur in different states of affairs in certain sorts of way, while members of P only recur in different states of affairs in certain other sorts of way, then this would warrant a partition of all possible constituents of states of affairs into two categories, which we could call, perhaps, 'universals' and 'particulars'. But what Ramsey's arguments suggest is that nothing like this is at all obviously the case.

---

[20] I have in mind, of course, the sort of ontology defended by David Armstrong in his *A World of States of Affairs* (Cambridge: Cambridge University Press, 1997).

7.5 THE ADVANTAGES OF A SUBSTANCE ONTOLOGY

However, it really should not surprise us, on reflection, that it is difficult to justify the universal–particular distinction within the framework of a states of affairs ontology. For that distinction first emerged in the context of a very different sort of ontology, namely, a *substance* ontology, Aristotle's being the prime exemplar. A substance ontologist certainly does not take states of affairs to be the basic building blocks of reality: rather, it is *substances*, of course, that are taken to have this status. And what is a substance? Well, that is a very long story, as long indeed as the history of western philosophy. But, without a doubt, one of the key ideas in the notion of substance is the notion of *ontological independence*. To be precise, this is not a single notion but a family of notions. However, it may be agreed by all substance ontologists that, in some suitable sense of 'ontologically independent', substances are ontologically independent entities. Spelling out the appropriate sense of 'ontologically independent' is no simple matter, but let us set aside this complication for the time being. What may then be said is that various *other* entities, whose existence we may need to acknowledge for one or other reason, may be held *not* to be ontologically independent entities in the same sense in which substances are, but rather to depend in some way for their existence or identity upon substances. And now the possibility emerges of constructing a system of ontological categories, with the distinguishing marks of any given category being the distinctive existence and identity conditions of its members.

For instance, it might be held that one class of entities is such that each member of the class depends for both its existence and its identity upon some specific substance, with many different members of the class depending in this way on the same substance, but no single member of the class depending in this way on two different substances. Call this class of entities 'Class 2', reserving for the class of substances the title 'Class 1'. What I have just said about Class 2 entities pretty much captures the ontologically distinctive features of what, in traditional substance ontologies, were called the 'modes' or 'individual accidents' of substances—the nearest equivalent in present-day metaphysics being so-called 'tropes' or, more exactly, monadic tropes. Again, it might be held that another class of entities—call it Class 3—is such that each member of the class depends for its existence, but *not* for its identity, upon a number of different substances. This pretty much captures the ontologically distinctive features of what are sometimes called 'substantial kinds'. Yet again, it might be held that still another class of entities—call it Class 4—is such that each member of the class depends for its existence, but not for its identity, upon a number of different entities belonging to Class 2. And this pretty much captures the ontologically distinctive features of what are sometimes called 'properties' or 'attributes', at least according to one metaphysical tradition (so-called 'immanent realism').

I don't pretend that the very brief sketch I have just given of how to construct a system of ontological categories in terms of various different ways in which members of the categories do or do not depend ontologically on the members of other categories is at all precise. The task of carrying out such a construction comprehensively and completely rigorously is a large and difficult one. Moreover, it is in principle possible to construct many different systems of categories in this way, so we are still left with the question of which of them we should prefer and on what grounds. But at present I am concerned only to defend the general approach, not any specific product of it—although, as the reader will know, my own preference is for a four-category ontology in which the four fundamental ontological categories correspond to something like the four classes just described. In section 7.8 below, however, I shall return to the issue and provide a rather more precise account of how I think this approach may be successfully implemented in the case of my own four-category ontology.

7.6 THE STATUS AND BASIS OF THE UNIVERSAL–PARTICULAR DISTINCTION

Now, we can see from the way in which the four classes were generated in the preceding section that there is a similarity between the way in which Class 3 entities depend ontologically upon Class 1 entities and the way in which Class 4 entities depend ontologically upon Class 2 entities: in each case, members of the first class depend for their existence, but not for their identity, upon a number of different members of the second class. This then entitles us to group Class 1 and Class 2 together and to do the same with Class 3 and Class 4. Entities belonging to the first group we can call 'particulars' and entities belonging to the second group we can call 'universals'. But it emerges, on this view, that there is not really an ontological *category* of particulars, nor one of universals, because there should be a commonality between the existence and identity conditions of the members of a genuine ontological category. The notions of 'particular' and 'universal' thus turn out strictly not to be categorial, but *transcategorial* notions—like the even more general notion of 'entity'.

Having introduced the distinction between universals and particulars in this way, we can now introduce some further terminology in which to talk of relationships between them. First, we need a word to describe the relationship between a universal and the particulars on which it depends ontologically and for this purpose we can use—as we have done in previous chapters—the word 'instantiation'. Second, we need a word to describe the relationship between a Class 2 entity and the Class 1 entity on which it depends ontologically and for this purpose we can, as previously, use the word 'characterization'. Third, we need a word to describe the relationship between a Class 4 entity and a Class 3 entity corresponding to the characterization relation between a Class 2 entity and a Class 1 entity. But for this purpose it seems

Ramsey's Problem and its Solution

again appropriate to use the word 'characterization'. Adopting the traditional terminology for entities belonging to the four different classes, we thus end up with a scheme according to which substances instantiate substantial kinds and are characterized by modes, while attributes characterize substantial kinds and are instantiated by modes—as is depicted in the diagram in Fig.7.1, which will by now be thoroughly familiar to readers of previous chapters.

As will be seen, to complete the picture I have included a third sort of relationship, between Class 1 entities and Class 3 entities, once again using for this purpose the word 'exemplification'.

However, a word of warning is in place concerning these three 'relationships', as I explained in Chapter 3 above. We are not to suppose that such relationships are themselves *entities* of any sort, to be found a place in one or other of our four fundamental ontological categories. This is the true lesson of Bradley's famous regress. It may be recalled here that Ramsey criticized W. E. Johnson's so-called characterizing 'ties' as being 'most mysterious objects'—and so they would be. But they should not be thought of as *objects*—that is, entities, or elements of being— at all. It may seem paradoxical to say this while at the same time speaking of 'them', for in so speaking we may appear to be reifying them as objects of reference. However, there is paradox here only on the assumption that nouns and pronouns are necessarily in the business of referring to entities—to parts of 'what there is'. But the task of describing reality and its structure cannot be exhausted by an inventory or list of what entities there are. The structure of being cannot just be a part of the being that is structured, or one of the beings that are so structured. The relationships between elements of being with which ontology is concerned are *formal* rather than *material* ones, as I tried to make clear in Chapter 3.

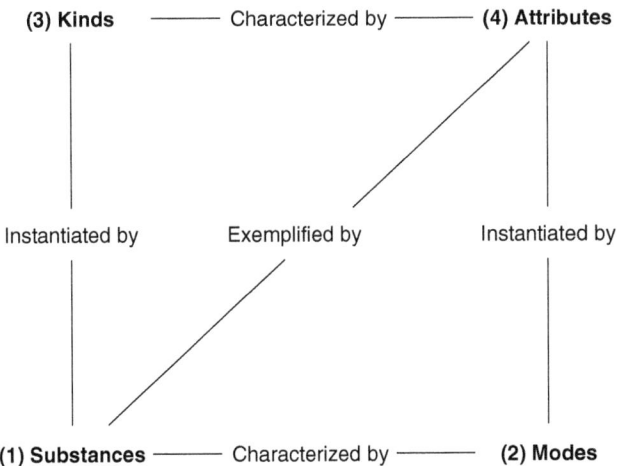

Figure 7.1. The four-category ontology

At this point it may be asked why it should not be possible to introduce the universal–particular distinction in something like the above fashion, but in the context of a states of affairs ontology. The answer is simply this. According to a states of affairs ontology, at most only atomic states of affairs may be ontologically independent entities, but certainly nothing else is: and the constituents of states of affairs are all ontologically dependent in exactly the same way on the states of affairs in which they occur. What this way is may be a matter for some dispute, but in the end I think that it will have to be acknowledged that a constituent of any state of affairs depends for its identity upon the *totality* of states of affairs—the 'possible world'—to which that state of affairs belongs. The consequence is that if a given constituent occurs in an actual state of affairs—a fact—then *it* does not exist in any other 'possible world', that is, it cannot be a constituent of any state of affairs belonging to a totality of states of affairs different from the actual one. The basic reason for this is that, as I mentioned earlier, constituents of states of affairs have to be thought of as 'abstractions from' or 'invariants across' states of affairs belonging to a totality of states of affairs—a 'possible world'. Since the very notion of a totality of states of affairs is the notion of a maximal collection of them, which cannot be added to, it does not make sense to think of something being an invariant across *more* than a single totality of states of affairs—a single 'possible world'. Invariants across states of affairs are 'world-bound' entities, depending for their identity on the world to which they are bound. Thus a states of affairs ontology induces either a very strong form of essentialism or, if one looks at it another way, a very strong form of anti-essentialism. More precisely, it undermines the sort of distinction between essence and accident which makes any interesting form of essentialism possible. This, again, should not surprise us: for the natural home of any interesting form of essentialism is a substance ontology.

7.7 THE DISSOLUTION OF RAMSEY'S PROBLEM

Let us now return, briefly, to Ramsey's example of Socrates and wisdom, looking at it from the perspective of a substance ontology—or, more specifically, from the perspective of the four-category ontology. From this perspective, Socrates is a substance—an ontologically independent entity—whereas wisdom is an attribute exemplified by Socrates. Socrates exemplifies this attribute in virtue of being characterized by a mode which is a particular instance of the atttribute. We can call this mode Socrates's wisdom, or the wisdom of Socrates. The attribute of which it is a mode—wisdom—in turn characterizes certain substantial kinds, such as, perhaps, the kind *philosopher*. This is a kind which Socrates himself instantiates. Hence the following statements turn out to be true: 'Socrates is wise', 'Philosophers are wise', and 'Socrates is a philosopher'. These are all subject/predicate sentences, grammatically speaking. But there is no simple mapping of the linguistic subject/predicate distinction onto any single ontological distinction, contrary to what Ramsey surmises in his diagnosis of the derivation of the universal–particular distinction. What, though, about Ramsey's contention that 'Socrates is wise' *means the same as* 'Wisdom is a characteristic of Socrates'? Well, it should be evident that the latter sentence is by no means a straightforward example of ordinary everyday English, in contrast with the former sentence. What precisely it should be taken to mean is a matter for dispute, because it uses an expression, 'characteristic', which is semi-technical in nature. It can, of course, simply be *stipulated* that 'Wisdom is a characteristic of Socrates' is to be understood as meaning the same as 'Socrates is wise', but then nothing of significance can be gathered from their equivalence. From the perspective of the four-category ontology, however, 'Wisdom is a characteristic of Socrates' appears to be open to two different interpretations. On the one hand, it might be taken to be saying that Socrates is characterized by a mode of the attribute of wisdom, which is tantamount to affirming the existence of Socrates's wisdom. On the other hand, it might be taken to be saying that Socrates instantiates a *kind* which is characterized by the attribute of wisdom, such as the kind *philosopher*. It emerges, then, concerning Ramsey's famous pair of sentences, that either they can just be stipulated to have the same meaning—in which case nothing interesting can be inferred from their equivalence—or else the second sentence of the pair has to be interpreted in the light of some specific account of ontological categories, in which case the account in question must be one which, like the four-category ontology, has been derived independently of the subject/predicate distinction. Consequently, Ramsey's example cannot in fact serve the purpose he attempted to put it to, that of impugning the credentials of the universal–particular distinction as being the product of the misunderstanding of a purely grammatical distinction.

What I have been trying to do in this chapter is to show how Ramsey's approach to the question of what grounds the universal–particular distinction has a result—his dismissal of that distinction as unfounded—that is entirely unsurprising, *given* his underlying assumption of a states of affairs ontology. But most of those who uphold the distinction should be unmoved, because it is one which really only makes sense in the very different environment of a substance ontology. Admittedly, if I am right, this leaves David Armstrong—the foremost current advocate of both a states of affairs ontology and the universal–particular distinction—in an awkward position. But he is big enough to fight his own battles and I would not presume to suggest how he might escape the difficulties that Ramsey's arguments present.

7.8 A RESPONSE TO MACBRIDE'S RAMSEIAN OBJECTIONS TO THE FOUR-CATEGORY ONTOLOGY

In a recent paper, Fraser MacBride has presented some interesting objections, with a distinctly Ramseian flavour, to my four-category ontology and to the place within it of the particular–universal distinction.²¹ Quite understandably, he has based his objections on a number of different publications of mine, appearing over a period of some fifteen years. Unsurprisingly, however, my views have not remained perfectly constant throughout that time but have undergone continual elaboration and revision, so that it must be difficult even for the most careful commentator to be quite sure about my current position on the matters in question. In this final section of the present chapter, therefore, I should like to further clarify my position with quite specific regard to the status and basis of the particular–universal distinction and to respond to MacBride's concerns.

The four-category ontology recognizes four basic ontological categories, which may be thought of as being generated by two basic and mutually independent formal ontological relations, *instantiation* and *characterization*, both of which are asymmetrical. As is depicted in Fig. 7.1 above, the categories are (1) *substances* or (more generally) individual *objects*, (2) relational and non-relational *modes*, (3) substantial or (more generally) objectual *kinds*, and (4) *attributes*—that is to say, *properties and relations*. Objects instantiate kinds and are characterized by modes, while properties and relations characterize kinds and are instantiated by modes. When I say that instantiation and characterization are *basic* formal ontological relations, I mean that they are not reducible to, or definable in terms of, other formal ontological relations (see Chapter 3 above). In this respect, I regard them as being akin to *identity*, which I also take to be a basic formal ontological relation. The four-category ontology is most perspicuously represented by *the Ontological Square*, Fig. 7.1 providing a version of this.

Within this ontological system, objects and modes are accorded the status of *particulars*, while kinds, properties and relations are accorded the status of *universals*. This is because the best way, in my view, to capture the traditional distinction between particulars and universals is by appeal to the *instantiation* relation. Universals are entities that are instantiated—that is, they have instances—while particulars are the entities that instantiate them. (I leave aside here the possibility of so-called 'higher-order' universals—universals that supposedly have other, 'lower-order' universals as their instances—because I am sceptical about the need to include them in our ontology.) Typically, universals are *multiply* instantiated— that is, they have many instances—but they need not be. At most I insist that every universal has at least one instance. (In other words, I espouse 'immanent' rather than 'transcendent' realism where universals are concerned.) Similarly, I consider that the traditional distinction between *subjects* and *predicables* is best captured by reference to the *characterization* relation. Subjects are entities that are characterized, while predicables are the entities that characterize them. However, I do not think that either *subjects* or *predicables* comprise a basic ontological

---

[21] See Fraser MacBride, 'Particulars, Modes and Universals: A Response to Lowe', *Dialectica* 58 (2004), pp. 317–33. Other objections have recently been raised by Max Kistler: see his 'Some Problems for Lowe's Four-Category Ontology', *Analysis* 64 (2004), pp. 146–51. I respond to these in my 'The Four-Category Ontology: Reply to Kistler', *Analysis* 64 (2004), pp. 152–7.

category. The notion of a subject is, in my opinion, essentially a *disjunctive* one: a subject is *either an object or a kind*. Likewise with the notion of a predicable. And likewise, too, with the notions of a particular and a universal. A particular is *either an object or a mode* and a universal is *either a kind or a property or relation*.

The Ramsey-inspired problem that Fraser MacBride raises for me is essentially the following. Given that the particular–universal distinction is to be drawn in the way that I propose, it is crucial that the four basic ontological categories that I invoke in drawing that distinction should themselves be unambiguously identifiable. Suppose, however, that we were to delete the terms denoting these categories in Fig. 7.1, leaving only the bracketed numbers to label the four corners of the Ontological Square, so that the four categories are effectively identified merely as 'category 1', 'category 2', 'category 3' and 'category 4'. And suppose, too, that we were to designate the vertical and horizontal sides of the square merely by the labels 'relation I' and 'relation C' respectively. It might then seem that we have been given insufficient information to distinguish each of the basic ontological categories unambiguously, because we have as yet no way of telling which side of the square is 'really' the top and which the 'bottom', nor which side is 'really' the left-hand side and which the right-hand side. To put it another way, it may seem arbitrary that we have labelled the bottom left-hand corner 'category 1' rather than 'category 2' or 'category 4', and similarly with respect to the other corners. For the same reason, then, it may be objected that in labelling the corners of the Ontological Square in my own preferred way, by such terms as 'objects', 'kinds', 'properties and relations' and 'modes', I have really done nothing to explain *how*, for example, objects differ from kinds or from modes, nor *how* properties and relations differ from modes or from objects. Hence, it may be complained, my utilization of these terms for the four basic ontological categories is essentially arbitrary and uninformative: for I have failed to explain what distinguishes these putative categories from one another. This, in essence, is the basis of MacBride's Ramsey-style objection to the four-category ontology.

Such an objection presumes that we have been offered, as yet, no insight into the meaning of the relational terms 'instantiation' and 'characterization' other than that they supposedly express different asymmetrical formal ontological relations capable of obtaining between entities belonging to different ontological categories. However, if someone pressing such an objection would only be satisfied by the provision of *reductive analyses* or *definitions* of instantiation and characterization, then I regret to say that I cannot satisfy such an objector, because I take the relations in question to be basic and irreducible. Moreover, since I believe that any adequate system of ontology must invoke at least some relations of this sort, I do not consider it to be a failing of my own system that it does so. Even so, this is not to imply that the four-category ontology contains no other resources with which to enrich our understanding of the basic categorial distinctions that it invokes. On the contrary—as I began to explain in section 7.5

above—such resources are available in the shape of the various *ontological dependence relations* that characteristically obtain between entities belonging to the different basic ontological categories. Three distinct ontological dependence relations are particularly relevant in this regard: *rigid* existential dependence, *non-rigid* existential dependence, and *identity dependence* (for details, see Chapter 3 above). Of these, the last is asymmetrical while the other two are not. And the most important of them for our present purposes are rigid existential dependence and identity dependence. (Here I note, incidentally, that MacBride makes no mention of identity dependence in his criticisms.)

It is characteristic of *objects*, I suggest, but also of *properties and relations*, that they are not identity-dependent upon entities belonging to other basic ontological categories. Thus, objects are only ever identity-dependent upon *other objects*. For example, a pile of rocks is identity-dependent upon the individual rocks that compose it. And some objects—most obviously, certain simple or non-composite ones—are not identity-dependent upon anything at all. Similarly, properties and relations are at most only identity-dependent upon *other properties and relations*. For example, conjunctive properties, if they exist, are identity-dependent upon the properties that are their conjuncts. In contrast, it is characteristic of *modes*, but also of *kinds*, that they are identity-dependent and also rigidly existentially dependent upon entities belonging to another ontological category. In the case of *modes*, the category in question is the category of *objects*—for modes are identity-dependent and also rigidly existentially dependent upon the objects that they characterize. For example, the particular redness of an individual red apple depends for its identity and rigidly for its existence upon that apple. In the case of *kinds*, the category in question is the category of *properties and relations*—for kinds are identity-dependent and also rigidly existentially dependent upon the essential properties that characterize them. For example, the kind *electron* depends for its identity and rigidly for its existence upon its essential properties of rest mass, charge and spin. So far, then, we have found a way of distinguishing, in terms of their characteristic dependency patterns, *objects* and *properties and relations*, on the one hand, from *modes* and *kinds* on the other. Finally, then, we note that the following further dependency patterns obtain. *Objects* are rigidly existentially dependent upon *kinds*, while *kinds* are only non-rigidly existentially dependent upon *objects*. For objects depend rigidly for their existence upon the 'highest' kinds that they instantiate, while all kinds depend non-rigidly for their existence upon the objects that instantiate them. Similarly, *modes* are rigidly existentially dependent upon *properties and relations*, while *properties and relations* are only non-rigidly existentially dependent upon *modes*. For modes depend rigidly for their existence upon the 'least determinate' properties and relations that they instantiate, while all properties and relations depend non-rigidly for their existence upon the modes that instantiate them. (I leave open here the question of whether or not we should in fact include 'determinable' as well as fully 'determinate' properties and relations in our ontology—a question that is, of course, different

Ramsey's Problem and its Solution   117

from that of whether or not we should include so-called 'higher-order' properties and relations.)

Let us now map these dependency patterns on to the Ontological Square, as follows (Fig. 7.2).

In this diagram, an arrow with a *solid* head (→) between a category **x** and a category **y** means that entities belonging to category **x** are identity-dependent and rigidly existentially dependent upon entities belonging to category **y**, whereas an arrow with a *V-shaped* head (→) between a category **x** and a category **y** means that entities belonging to category **x** are rigidly existentially dependent upon entities belonging to category **y** while entities belonging to category **y** are only non-rigidly existentially dependent upon entities belonging to category **x**. Thus, both arrows represent *asymmetrical* relations, the second obviously so and the first in virtue of the asymmetry of identity dependence. Moreover, the four arrows taken together serve to identify unambiguously the four corners of the square, in terms of the number and kinds of arrow-heads found at each corner. Thus, the bottom left-hand corner has just one solid arrow-head located at it, the top left-hand corner has just one V-shaped arrow-head located at it, the top right-hand corner has one solid and one V-shaped arrow-head located at it, and the bottom right-hand corner has no arrow-heads located at it.

Of course, the foregoing claim—that the four corners of the Ontological Square are unambiguously identifiable in precisely the way I suggest—would be undermined if it could be successfully argued that I am mistaken in my assessment of the dependency patterns obtaining between entities belonging to the various categories. For example, some ontologists may urge that *objects* are identity-dependent upon *haecceities*, the latter conceived as belonging to the category of *properties and relations*. To this kind of worry, which MacBride himself articulates, I reply as follows. First, those who are inclined to raise such an objection are already committed to the sorts of categorial distinction that I am trying to defend and so are in no position to put it to sceptical use in attacking those very distinctions. Second, it is perfectly conceivable that, even if the particular pattern

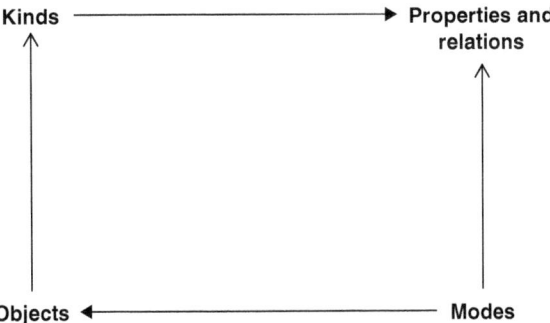

Figure 7.2. Patterns of ontological dependency between the four basic categories

of dependencies that I appeal to should turn out to be mistaken in some respects, another pattern could be defended which would serve the same purpose—that is, which would serve to identify unambiguously the four corners of the Ontological Square. My relatively modest aim in this section has been to show that the prospects for defeating a Ramsey-style objection to the four-category ontology and the particular–universal distinction that it embodies are considerably more promising than MacBride suggests.

PART III

METAPHYSICS AND NATURAL SCIENCE

8

Dispositions and Natural Laws

8.1 DISPOSITIONAL VERSUS CATEGORICAL PROPERTIES

Many philosophers like to distinguish between 'dispositional' and 'categorical' *properties*, but then puzzle over the relation between them. Some of these philosophers maintain that all dispositional properties have categorical 'bases'—that is, that dispositional properties are always 'grounded in', 'realised by', or 'supervene upon' categorical properties of their bearers, or categorical properties and relations of the microstructural constituents of their bearers. Some of these same philosophers go on to argue that dispositional properties are 'second-order' properties—that is, that a dispositional property is the property of having some first-order categorical property in virtue of which its bearer is disposed to behave in some specific way in suitable circumstances.[1] Other philosophers contend that there can be 'pure' dispositional properties or powers, which have no categorical basis at all, or even, more radically, that *all* properties are dispositional.[2] Yet others (notably, C. B. Martin) maintain—or, at least, have been interpreted as maintaining—that all properties have both a dispositional and a categorical (or 'qualitative') *aspect* or *side*, so that it is an error to suppose that the dispositional–categorical distinction is a distinction between *types of property*.[3]

---

[1] For a view of this kind, see Elizabeth Prior, *Dispositions* (Aberdeen: Aberdeen University Press, 1985). See also Elizabeth Prior, Robert Pargetter and Frank Jackson, 'Three Theses about Dispositions', *American Philosophical Quarterly* 19 (1982), pp. 251–7. The standard work on the subject of dispositions at present is undoubtedly Stephen Mumford's *Dispositions* (Oxford: Oxford University Press, 1998).

[2] See D. H. Mellor, 'In Defense of Dispositions', *Philosophical Review* 83 (1974), pp. 157–81.

[3] See, especially, Martin's contributions to D. M. Armstrong, C. B. Martin and U. T. Place, *Dispositions: A Debate*, ed. Tim Crane (London: Routledge, 1996), for example at p. 133. See also C. B. Martin and John Heil, 'The Ontological Turn', *Midwest Studies in Philosophy* XXIII (1999), pp. 34–60. I should stress that most recently Heil and Martin have distanced themselves emphatically from all talk of 'aspects' in favour of an account which simply proclaims the *identity* of the dispositional and the qualitative: see John Heil, *From an Ontological Point of View* (Oxford: Clarendon Press, 2003), ch. 11. See also note 19 below. Stephen Mumford holds a similar position, maintaining that 'The dispositional and the categorical do not ascribe two different types of property . . . rather, they are two modes of denoting the very same properties' (*Dispositions*, p. vi).

I am not persuaded that any of these views is correct, but of all of them I find the last the least implausible, especially insofar as it denies that the dispositional–categorical distinction is properly to be construed as a distinction between types of property. I should also say, however, that I do not particularly favour using the term 'categorical' to express the distinction at issue, preferring the term 'occurrent', for reasons which will emerge in due course.

8.2 PREDICATES AND PROPERTIES

It is a familiar—but still insufficiently emphasized—point that not every meaningful predicate expresses a real (that is, an existing) property. We know this as a matter of logic, because the predicate 'is non-self-exemplifying' is perfectly meaningful—and, indeed, is truly applicable to some things—and yet there cannot, on pain of contradiction, be such a property as the property of being non-self-exemplifying: because if such a property exists, it must either exemplify itself or not exemplify itself—and if it does, then it doesn't, and if it doesn't, then it does. Either way, then, we have a contradiction: so such a property does not exist. This, of course, is just a version of Russell's paradox. But given that not every meaningful predicate expresses an existing property, we need a way of determining, if possible, when precisely it is that a meaningful predicate does express an existing property. In other words, we need an answer to the following question: when is it that we are entitled to say that the property of being F exists, where 'F' is some meaningful predicate? This is a far from easy question to answer.

Elsewhere, I have tried to defend the following answer to this question: the property of being F exists just in case (i) there is some property which is exemplified by all and only those things which are F and (ii) there is something which is F.[4] There is nothing circular about this proposal. It is true that it specifies the existence conditions of the property of being F in terms which involve quantification over properties. But that is unexceptionable, because my aim is not to state what it is for properties in general to exist—only what it is for there to exist the property of being F, where 'F' is a quite specific meaningful predicate. Satisfaction of these existence conditions is a far from trivial matter—and it certainly does not appear that the proposal commits us to the absurdity of supposing that every meaningful predicate expresses an existing property. The idea behind the proposal is the seemingly common-sense one that the property of being F is what all and only the Fs have in common, if indeed they do all have something in common. The proposal allows room for the possibility that in many

---

[4] See my 'Abstraction, Properties, and Immanent Realism', *Proceedings of the Twentieth World Congress of Philosophy, Volume 2: Metaphysics*, ed. Tom Rockmore (Bowling Green, OH: Philosophy Documentation Center, 1999), pp. 195–205.

cases the *F*s do *not* have something exclusively in common—for instance, that there is nothing that all and only *games* have in common. I should emphasize that I am, of course, talking here about properties in the sense of *universals*, not in the sense of property-instances, 'tropes', or (my own preferred term) *modes*. I should also remark that the foregoing proposal will need some adjustment in order to accommodate the distinction that I shall be making shortly between dispositional and occurrent *predication*. But before coming to that I want to turn to some more general matters of ontology.

8.3 THE FOUR-CATEGORY ONTOLOGY: A BRIEF RÉSUMÉ

Some metaphysicians favour sparse ontologies where properties are concerned—some implausibly denying the existence of properties altogether, others only acknowledging the existence of properties in the sense of *universals*, and yet others only acknowledging the existence of properties in the sense of tropes or modes (that is, properties as *particulars*, or 'particularized' properties). Some metaphysicians are equally sparing concerning the bearers of properties, that is, concerning *substances* or 'objects'. Some of them maintain that objects are just bundles of properties, either bundles of universals or bundles of tropes. My own position is more liberal, both with regard to properties and with regard to substances. As readers of earlier chapters will know, I favour a 'four-category' ontology, which seems also to have been favoured by Aristotle, at least in the *Categories*. According to this ontology, there are both universals and particulars. Particulars fall into two distinct categories: individual substances on the one hand and modes (or tropes) on the other, with substances being (in a certain sense which I have defined elsewhere) *ontologically independent* entities while modes are ontologically dependent upon the substances which are their 'bearers'.[5] Equally, universals fall into two distinct categories: substantial universals, or *kinds*, and non-substantial universals, or *properties*. (I ignore relations for the time being, for the sake of simplicity, but in fact I take these to be non-substantial universals whose particular instances are relational tropes.) Individual substances are particular instances of substantial universals, or kinds, while modes are particular instances of non-substantial universals, or properties. We shall see shortly how the four-category ontology bears upon the question of the nature of dispositions and their relation to laws.

---

[5] For more on the notion of ontological dependence in play here, see my *The Possibility of Metaphysics: Substance, Identity, and Time* (Oxford: Clarendon Press, 1998), ch. 6.

8.4 DISPOSITIONAL VERSUS OCCURRENT PREDICATION

I have already indicated that I am opposed to drawing a distinction between dispositional and categorical (or occurrent) *properties*, as though what is at issue here is a distinction between types of property. Rather, I want to distinguish between dispositional and (as I prefer to call it) occurrent *predication*.[6] And, as I have already remarked, not every predicate expresses a corresponding existent property. The distinction in question is exhibited in such pairs of sentences as 'This piece of rubber *stretches* (or *is stretchy*)' and 'This piece of rubber *is stretching*', or 'This stuff *dissolves* in water' and 'This stuff *is dissolving* in water'. Now, my view is that, even when a predicate *does* express a real property, it expresses *one and the same* property (in the sense of *universal*) irrespective of whether the predication involved is dispositional or occurrent in character. But before I try to explain what is behind this suggestion, let us look at some further examples.

Consider, then, properties of shape and colour, which many philosophers regard as paradigm examples of categorical and dispositional properties respectively. Thus, for example, the property of being square is often held to be a paradigmatically categorical property, while the property of being red is often held to be a dispositional property—the property of being disposed to induce 'red' sensations in normal percipients in normal viewing conditions, or something like that. (Here I am setting aside the deeper question of whether colour properties really exist; let us suppose for present purposes that they do.) But I would urge that each of the predicates 'is square' and 'is red' has both a dispositional and an occurrent interpretation. Thus, I suggest, a surface which 'is red' in the *dispositional* sense is one which, nevertheless, is *not* red, in the *occurrent* sense, in a darkened room or under blue light: in those circumstances, I suggest, the surface is grey or black in the occurrent sense. If we used verbs instead of adjectives to express colour, as some languages do, this would be more obvious. We could then render more explicit the distinction between the occurrent and dispositional senses of our ambiguous colour predicate 'is red' by saying that a surface which is not *redding* under blue light may none the less be a *reddy* surface—just as we say that a piece of rubber which is not *stretching* may none the less be *stretchy*. Equally, I think that a surface can be both occurrently and dispositionally *square*—but, again, that this should not be conceived of as a distinction between different types of property of the surface. For example, a rubber eraser may be 'square' in the dispositional sense while also being 'trapezoid', say, in the occurrent sense, when it

---

[6] See further my *Kinds of Being: A Study of Individuation, Identity and the Logic of Sortal Terms* (Oxford: Blackwell, 1989), chs. 8–10, much of which is based on my earlier papers 'Sortal Terms and Natural Laws', *American Philosophical Quarterly* 17 (1980), pp. 253–60 and 'Laws, Dispositions and Sortal Logic', *American Philosophical Quarterly* 19 (1982), pp. 41–50.

is subjected to certain distorting stresses. (Note, however, that I don't want to say that a surface is occurrently non-square simply when it *looks* non-square to some observer, because I take it that squareness is a primary property whose possession by an object has nothing to do with its relation to observers.)

8.5 THE ONTOLOGICAL GROUND OF THE DISPOSITIONAL–OCCURRENT DISTINCTION

I have said that I don't regard the dispositional–occurrent distinction to be one between types of property, but this is not to say that I do not think that it has any ontological ground at all. Quite the contrary. My proposal is simply this. *Occurrent* predication involves the attribution to an object of some *mode* of a property: that is, it involves the attribution to an object of a property-instance or trope which instantiates some property (in the sense of *universal*). By contrast, *dispositional* predication involves the attribution of some property (in the sense of *universal*) to an object's substantial *kind*. Thus, for example, to say that a certain piece of common salt is water-soluble is, on my view, to say that this object is something of a water-soluble kind. This is a proposal which is reminiscent, incidentally, of one which W. V. Quine has advanced, although only as part of a general attempt to downgrade the scientific significance of our talk of dispositions—an attempt with which I have no sympathy.[7]

More precisely, the proposal is this. A sentence of the form '*a* is occurrently *F*' means '*a* possesses a mode of *F*ness', whereas a sentence of form '*a* is dispositionally *F*' means '*a* instantiates a kind *K* which possesses *F*ness'. Thus, according to this view, properties (in the sense of *universals*) primarily characterize *kinds* and only derivatively or indirectly characterize individual substances or objects. Properties, however, can derivatively or indirectly characterize individual objects in either of two quite different ways. One way in which properties can indirectly characterize individual objects is inasmuch as those objects possess particular instances of those properties, that is, inasmuch as they possess modes which instantiate those properties. The other way in which properties can indirectly characterize individual objects is inasmuch as those objects belong to kinds which are themselves characterized by those properties. The first type of case is what is at issue when we use 'occurrent' predication and the second type of case what is at issue when we use 'dispositional' predication.[8]

---

[7] See W. V. Quine, 'Natural Kinds', in his *Ontological Relativity and Other Essays* (New York: Columbia University Press, 1969).
[8] I should remark, incidentally, that the origins of this proposal are to be found in the last three chapters of my *Kinds of Being* (see especially pp. 170–1), although at the time of writing that book I was not convinced about the existence of modes (property-instances or tropes).

The following—by now very familiar—diagram (Fig. 8.1) depicts my proposed conception of the ontological ground of the dispositional–occurrent distinction.

From Fig. 8.1 it should be immediately apparent that, as I have just urged, individual substances, or objects, may be said to be *indirectly* characterized by non-substantial universals, or properties, in two different ways—namely, either via their instantiation of substantial kinds which are characterized by those properties (a route depicted in the diagram by the left-hand vertical line followed by the upper horizontal line), or else via their characterization by property-instances, or modes, which instantiate those properties (a route depicted in the diagram by the lower horizontal line followed by the right-hand vertical line). According to my proposal, the first of these ways is expressed by a sentence whose subject-term refers to a particular object and whose predicate is dispositional, while the second of these ways is expressed by a sentence whose subject-term refers to a particular object and whose predicate is occurrent. Given that the ontology is as I have depicted it, it is only to be expected that natural language should provide some systematic means of representing these two different types of circumstance—and my suggestion is that the distinction between dispositional and occurrent predication is precisely one such means that is provided by English and many other natural languages.

Of course, it is important to understand correctly what logical relations do and do not obtain between sentences involving dispositional and occurrent predications of the same property. In particular, it is clear that we must not say that a sentence of the form '*a* is dispositionally *F*' entails one of the form '*a* is occurrently *F*'. Given my analysis of what such sentences mean, I must clearly hold that '*a* instantiates a kind *K* which possesses *F*ness' does not entail '*a* possesses a mode of *F*ness'. But

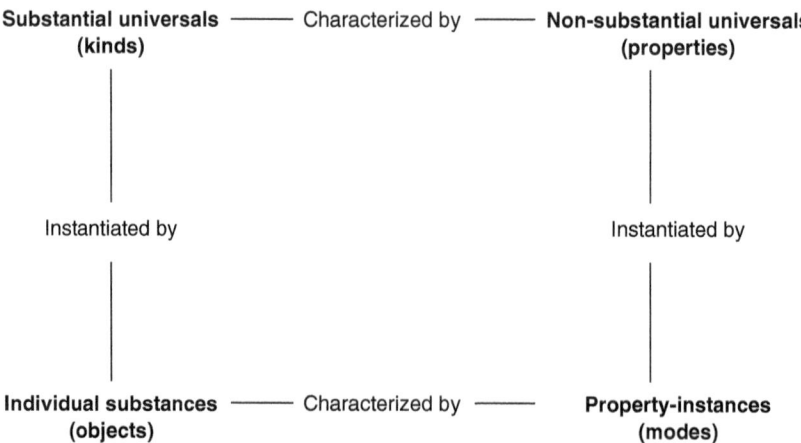

Figure 8.1. The ontological ground of the dispositional–occurrent distinction

this is perfectly consistent with what I shall be going on to say shortly about laws of nature.[9]

Just before I say something about laws, however, I want to extend my use of the expressions 'dispositional' and 'occurrent' in order to make the following distinction between two fundamentally different types of *states of affairs* involving individual objects. I suggest that we may describe an object's possessing a mode of some property (universal) as being an *occurrent* state of affairs, whereas an object's belonging to a kind which possesses some property (universal) may correspondingly be described as being a *dispositional* state of affairs.

8.6 LAWS OF NATURE

We are now left with a further important type of state of affairs still to be identified: one which simply consists in some kind's possessing some property (universal). But we already have a name for this type of state of affairs: it is a *law of nature*. (Quite literally, it is a law concerning the *nature* of things of the kind in question.) For example: common salt's being water-soluble is such a law, as is rubber's being elastic (or 'stretchy'). When we say that a *particular* piece of rubber 'is elastic'—that it 'stretches', or 'is stretchy'—what we are saying, according to my account, is that this object belongs to a kind, K, such that the *law* obtains that K is elastic or 'stretches'. In this case, of course, the kind in question is *rubber*. Similarly, when we say that a particular object is 'water-soluble', what we are saying is that it belongs to a kind, K, such that the law obtains that water dissolves K (or, in other words, that K is water-soluble). Thus, the kind in question might be common salt, or sodium chloride. In other words, on this view, individual objects possess their various natural 'powers' in virtue of belonging to substantial kinds which are subject to appropriate laws—these laws consisting in the possession by such kinds of certain properties (in the sense of universals), or in the standing of kinds in certain relations to one another. And notice that, in expressing laws themselves, we naturally have recourse to *dispositional* predication, asserting such sentences as 'Rubber stretches' ('Rubber is elastic'), 'Water dissolves common salt' ('Common salt is water-soluble') and 'Planets move in elliptical orbits'. But this is now wholly understandable, given that dispositional predication is a grammatical device for signalling the attribution of properties in the sense of *universals* (rather than in the sense of modes or tropes) and that laws consist in the possession by kinds of properties in this sense.[10]

---

[9] I develop a logic of dispositional and occurrent predication, which is demonstrably consistent, in my *Kinds of Being*: see the appendix to ch. 9. There are, however, certain further complications, discussed in ch. 10 of that book, which I am ignoring for present purposes.

[10] That the canonical form of a statement of natural law involves dispositional predication and a substantial kind term in subject position is something that I argue for at length in my *Kinds of Being*, ch. 8.

I shall say more about this view of laws later, contrasting it with a more familiar view of laws which likewise makes reference to universals, namely, David Armstrong's view. Incidentally, while I happily talk here about 'states of affairs', I do not (unlike Armstrong, for instance), want to make states of affairs the basic building blocks of my ontology: for me these are constituted, rather, by entities belonging to my four basic categories—substantial and non-substantial particulars, and substantial and non-substantial universals.[11]

Before moving on I want to deal with a further question which may very naturally be raised at this point. According to my account of the distinction between dispositional and occurrent predication, the distinction corresponds—where *individual substances* or *objects* are the subjects of predication—to two different ways in which such objects may be 'indirectly' characterized by non-substantial universals. In terms of Fig. 8.1 above, these two different ways are represented by two different routes around the square—*the Ontological Square*, as I have called it in previous chapters—from the bottom left-hand corner to the top right-hand corner. I have also said, however, that dispositional predication is involved when non-substantial universals are predicated of *substantial universals* or *kinds*—such predications being expressive of laws—and this sort of case is, clearly, represented in Fig. 8.1 by the *direct* route from the top left-hand corner to the top right-hand corner of the square. The question which naturally arises at this point is the following, however. Should we not expect there also to be a class of *occurrent* predications concerning substantial universals or kinds, represented by the *indirect* route between the top left-hand corner and the top right-hand corner of the square? Indeed we should—and our expectations will not be disappointed. Consider again one of my previous examples of a *dispositional* predication concerning a kind, 'Rubber *stretches*', and compare this with the statement 'Rubber *is stretching*', in which the predicate clearly has *occurrent* force. What can the latter statement plausibly be taken to mean? Just this, I suggest: that some individual piece or portion of rubber—that is, an *individual substance* or *object* which is an instance of the kind *rubber*—*is stretching*.[12] But this is, indeed, precisely to say that the kind in question is 'indirectly' characterized by the non-substantial universal in question in virtue of the kind possessing an instance which is characterized by a mode of that non-substantial universal—a relationship depicted in Fig. 8.1 by the *longer* route around the square from the top left-hand corner to the top right-hand corner. The fact that natural language includes occurrent predications of this sort which can be accommodated in this simple fashion by the theory now being advanced constitutes, I believe, further evidence in support of the theory. But now let me return to my previous theme—the natural powers of individual objects.

---

[11] For Armstrong's position, see his *A World of States of Affairs* (Cambridge: Cambridge University Press, 1997). [12] See further my *Kinds of Being*, p. 145 and pp. 167–9.

8.7 NATURAL POWERS AND CONDITIONAL STATEMENTS

The natural powers of individual objects can be either relational or non-relational in character. For example, water-solubility is a relational power, whereas flammability is a non-relational power. The possession by an individual object of a relational power implies the existence, or at least the possible existence, of individual objects possessing a reciprocal power. For instance, corresponding to a particular piece of common salt's power of water-solubility (its 'liability' to be dissolved by water) is the reciprocal power of any particular quantity of water to dissolve common salt. By contrast, the possession by an individual object of a non-relational power, such as flammability, has no such implication concerning the powers of other objects. This distinction has at its ontological root the distinction between relational and non-relational *laws*. The law that water dissolves common salt is a relational law, since it consists in the obtaining of a certain relation (in the sense of a *universal*, of course) between the substantial kinds water and common salt. The law that benzene is flammable is a non-relational law, consisting in the possession of a monadic property (universal) by a single substantial kind, benzene. This is perhaps more obvious if we state that law in the form 'Benzene burns' rather than in the form 'Benzene is flammable'—for then the point emerges in the grammatical fact that 'burns' in this context is an intransitive rather than a transitive verb.

Another important distinction between natural powers, which is intimately related to but should not be conflated with that between relational and non-relational powers, is the distinction between *conditional* and *non-conditional* powers. A conditional power is a power to behave thus-and-so *if*, or *when*, such-and-such circumstances or conditions arise. Any object which possesses a relational power thereby also possesses a species of conditional power. For instance, because a piece of common salt has the relational power (or liability) to be dissolved by water, it has the conditional power of dissolving *if*, or *when*, it is immersed in water. Generalizing, we may say that a conditional power is a power to F if (or when) G, whereas a non-conditional power is a power to F *simpliciter*.[13] This distinction is, again, ontologically rooted in a distinction between conditional and non-conditional *laws*. That radium undergoes spontaneous decay is a non-conditional law, whereas that common salt dissolves if (or when) it is immersed in water is a conditional law. Notice here that the law that benzene is flammable, properly understood, is not a non-conditional law, since benzene does not undergo combustion spontaneously: the law is, rather, that benzene burns, or undergoes combustion, *if*, or *when*, it is subjected to a flame.

---

[13] For further discussion of the logical distinction at issue here, see my *Kinds of Being*, ch. 10.

It seems that conditional powers are amongst the 'truthmakers' of many conditional statements—though by no means of all conditional statements (not, for instance, those whose truth can be established by logical or mathematical proof).[14] Conditional power statements, however, certainly cannot be *analysed* in terms of conditional statements, for familiar reasons (some of which have been clearly laid out by C. B. Martin).[15] Neither the conditional power statement 'This piece of common salt has the power (or liability) to dissolve if (or when) it is immersed in water' nor the relational power statement (or disposition statement) 'This piece of common salt is water-soluble' is analysable as meaning anything like 'If this piece of common salt is (or were to be) immersed in water, then it is (or would be) dissolving'.

It is a familiar point that there are all sorts of ways—indeed, indefinitely and perhaps even infinitely many different ways—in which the 'manifestation' of a conditional power may be blocked or prevented, even when the 'condition' appropriate to the manifestation of the power obtains. Common salt will fail to dissolve in water if the water happens already to be saturated by common salt, for instance. Clearly, it is pretty well vacuous to say that a piece of common salt will dissolve if it is immersed in water *and nothing prevents its dissolution by the water*. We are simply in no position to list all the possible preventive factors in virtue of which a piece of common salt would fail to dissolve even if it were immersed in water: consequently, when we describe a piece of common salt as being 'water-soluble', we cannot possibly *mean* by this that a certain complex conditional statement, explicitly excluding the presence of all those possible preventive factors, is true—for we are unavoidably incapable of formulating an appropriate antecedent for any such conditional statement. As I have already indicated, however, this need not preclude us from saying that, on a particular occasion, an object's possession of a certain relational or conditional power was at least *part of* the 'truthmaker' of a certain conditional statement. If it was true to assert, on a given occasion, the counterfactual conditional sentence 'If this piece of common salt had been immersed in that water, then it would have dissolved', part of what will have made this conditional sentence true is the fact that the salt possessed a power to be dissolved by water and the water possessed a reciprocal power to dissolve common salt.

8.8 LAWS AND UNIVERSALS

I agree with those philosophers, such as David Armstrong, who maintain that natural laws involve universals, rather than consisting in regularities or constant

---

[14] On the notion of a 'truthmaker', see Kevin Mulligan, Peter Simons and Barry Smith, 'Truth-makers', *Philosophy and Phenomenological Research* 44 (1984), pp. 287–321, and Chapters 11 and 12 below.

[15] See C. B. Martin, 'Dispositions and Conditionals', *Philosophical Quarterly* 44 (1994), pp. 1–8. See also George Molnar, *Powers: A Study in Metaphysics* (Oxford: Oxford University Press, 2003), ch. 4.

conjunctions amongst particulars.[16] The law that water dissolves common salt does not consist in the general state of affairs that every particular piece of common salt that was, is, or will be immersed in any particular quantity of water was, is, or will be dissolved by that water, even if that general state of affairs does as a matter of fact obtain—although, of course, it does not in fact obtain, because on many occasions various preventive factors have blocked the dissolution of some salt by some water in which it has been immersed. But, partly for this very reason, nor do I think that the law in question consists in the holding of a second-order relational universal of nomic or natural necessitation between first-order universals which somehow implies that such a general state of affairs, or constant conjunction, obtains. On Armstrong's account of laws, a law is a state of affairs which consists in one universal's standing in a relation of necessitation to another, that is, which consists in *F*ness necessitating *G*ness, for certain properties (first-order universals) *F*ness and *G*ness—the implication supposedly being that, because such a state of affairs obtains, so also does the general state of affairs of every particular thing which is *F* also being a thing which is *G*. (Of course, not all laws will have precisely this form on the Armstrong model, which also allows, in any case, for probabilistic laws. But I am focusing on the simplest case for ease of exposition: what I have to say about that can easily be generalized to apply to more complex cases.)

By my own account of laws, in contrast, laws (again in the simplest case) consist in the characterization of substantial universals (kinds) by non-substantial universals (properties), or in the holding of relations (that is, relational universals) between two or more substantial universals. For example, *benzene* is characterized by *burning* and *water* and *common salt* are related by *dissolving*. It is in precisely such terms, after all, that we actually express these laws: we say that *benzene burns* (benzene is flammable) and that *water dissolves common salt* (common salt is water-soluble). So, on this view, while laws do indeed involve universals, and sometimes involve relations between universals, they do not involve some special 'second-order' relation of necessitation. Nor, on my view, do laws imply the existence of constant conjunctions amongst occurrent states of affairs involving particulars. The law that water dissolves common salt does not, for instance, imply that every quantity of water that is in contact with some piece of common salt is dissolving that piece of salt, or anything like that. Laws, in general, only have implications for how particular objects *tend*, or *are disposed*, to behave in various circumstances, not for how they actually *do* behave.[17]

Armstrong, it seems, is compelled to appeal to a second-order relation of necessitation to link the universals involved in a law because he does not distinguish—as I believe one must—between substantial and non-substantial universals.[18] In effect, for Armstrong, all first-order universals are non-substantial

---

[16] For Armstrong's view, see especially his *What is a Law of Nature?* (Cambridge: Cambridge University Press, 1983).
[17] See further my 'What *is* the "Problem of Induction"?', *Philosophy* 62 (1987), pp. 325–40.
[18] See Armstrong, *A World of States of Affairs*, pp. 65–8.

universals (either properties or relations), which is why, for him, a law (in the simplest case) can only consist in one such universal, *F*ness, necessitating (or probabilifying) another, *G*ness. But because I, in contrast, recognize both substantial and non-substantial universals, I can say that what 'links' the universals involved in a law is not some mysterious second-order relation, but simply the familiar characterizing or predicative tie, which 'links' any property to its 'bearer'. (I don't say that the nature of the predicative tie is completely unproblematic—far from it—but since Armstrong is as much committed to it as I am, it is an advantage of my view of laws over his that I am committed only to this and not also to some mysterious second-order relation of necessitation between universals.) The form of a law, in the simplest case, is just this, on my view: substantial kind *K* is characterized by *F*ness, or, even more simply, *K* is *F*.

8.9 NOMIC NECESSITY

What, then, about 'nomic necessity'? If, by this, is meant 'lawlike connection', then my view is that the term is a misnomer. Laws of nature are not—or, at least, not always—necessary states of affairs, but contingent ones. It is true that what belongs to the *essence* of a kind, *K*, attaches of necessity to all individual objects of kind *K*. (For instance, it is part of the essence of any kind of material object that it is extended in space.) But here we are talking of genuine, *metaphysical* necessity. So-called 'natural' or 'physical' necessity is at best a species of 'relative' necessity: a matter of what is necessarily the case *given* that some contingent truth obtains. Here I should emphasize that it is not my view that the natural laws concerning a kind *K* involve all and only those properties which belong to *essence* of *K*, in the metaphysical sense of 'essence'. Thus, in my view, it is not part of the essence of water that it dissolves common salt. I believe that water—that very substance—could, very arguably, exist in a possible world in which there was not a natural law that water dissolves common salt. (Here I use the language of 'possible worlds' without any presumption of realism concerning possible worlds.) At most we can say that *if* there is a law, in a given possible world, that water dissolves common salt, then it follows of necessity in that world that any particular quantity of water has a tendency or disposition to dissolve any piece of common salt with which it may come into contact. A 'lawlike connection' is simply the sort of connection between universals that we find exhibited in laws of nature—and this, as I have said, is simply the familiar characterizing or predicative tie, not some mysterious relation of contingent necessity, which is somehow projected upon particular instances of the universals involved, as in Armstrong's account. (So what *is* the 'essence' of water, say, given that it is not part of the essence of water that it dissolves common salt? I am happy to say that it is part of the essence of water that it consists of hydrogen and oxygen atoms in the ratio of two to one; but the point is that I consider that the laws governing hydrogen and oxygen atoms are quite conceivably different in different possible worlds—that these laws are not metaphysically necessary states of affairs, but merely contingent ones.)

Of course, one reason, historically, why some philosophers have believed in 'nomic necessity' is that they imagine that some such notion is needed in order to distinguish between lawlike statements and mere accidental generalizations—with the former being seen as carrying an implication of necessity or 'necessary connection'. But I regard this motivation as being confused. Statements of natural law should not be seen as modalized versions of universally quantified statements concerning particulars, of the form (in the simplest case) 'Necessarily, for any individual, x, if x is F, then x is G' (where the modality in play is somehow 'weaker' than metaphysical or logical necessity). Rather, statements of law are statements of the form (again, in the simplest case) 'K is F', where 'K' is a substantial kind term (or 'Ks are F', if 'K' is a sortal term rather than a mass term). Such a statement is not, however, analysable in terms of any logical operation upon a universally quantified statement concerning particulars, such as 'For any individual, x, if x is (a) K, then x is F', for reasons which I have sufficiently explained elsewhere.[19] Consequently, it is neither necessary nor possible to distinguish between lawlike statements and accidental generalizations in the way in which many philosophers have imagined that we can or should: the distinction between such statements is clear enough already, once we correctly apprehend their respective logical forms. The appeal to 'nomic necessity', it seems, is partly the product of a misguided attempt to overcome what was correctly perceived to be the inadequacy of the constant conjunction or regularity view of laws. However, these are matters that I shall discuss in more detail in Chapters 9 and 10 below.

8.10 A COMPARISON WITH MARTIN'S ACCOUNT

I said earlier that I sympathize with C. B. Martin's claim that there are not two different types of property—dispositional and categorical (or occurrent) properties—but that I am not happy with the suggestion that all properties have both a dispositional and a categorical (or 'qualitative') *aspect* or *side*. I do not really understand what could be meant by an 'aspect' of a property in this sense, given that we obviously cannot be supposed to be talking about some sort of 'second-order' property—a property of a property.[20] My preference, as I have explained above, is to say instead that the dispositional–occurrent distinction is one between

---

[19] See my *Kinds of Being*, ch. 8.
[20] John Heil makes it clear that in his view second-order properties are *not* at issue here, and indeed repudiates talk of 'aspects' in this context: see his *From an Ontological Point of View*, pp. 118–20. What he does say, though, is that 'A property's dispositionality and its qualitativity are, as Locke might have put it, the selfsame property differently considered' (p. 112). However, that is pretty much how so-called 'dual-aspect' theories of the mind–body relation (such as Spinoza's) have traditionally been presented, namely, as versions of the *identity* theory—so it is not clear to me why Heil and Martin should feel it necessary to distance themselves from talk of 'aspects' altogether.

ways of *predicating* properties, but one which is ontologically grounded in the two different ways in which particular objects can be 'indirectly' characterized by universals. Clearly, Martin must reject anything like this account, because he rejects universals altogether: he favours a two-category ontology of individual substances and particular properties or tropes (what I call 'modes'). Hence, given that he holds—rightly in my view—that there are not two different types of property, dispositional and categorical, and yet wants to retain that distinction in some form, he is compelled to say that every particular property or trope is at once dispositional and categorical (or qualitative) in nature. But, as I say, I do not really understand what this could *mean*.

I prefer to say that every state of affairs which consists in the possession of some trope or mode by a particular object is an *occurrent* state of affairs, pure and simple, and that a *dispositional* state of affairs is one which consists in some particular object's instantiating a substantial kind which is characterized by some property (in the sense of universal, of course). It seems to me that my account is less mysterious than Martin's, though I concede that this will seem so only to someone who does not find the notion of universals itself mysterious. However, judging by the history of philosophy, very many philosophers would appear to fall into this category. For even those philosophers who deny the existence of universals generally concede, unsurprisingly, that they comprehend the notion of what it is whose existence they are denying. So, even for these philosophers, my account should appear at least comprehensible.

On the other hand, I cannot deny that Martin's ontology is simpler than mine, involving as it does two basic categories instead of four. So it is incumbent upon me to justify my more liberal ontology. Why, then, do I admit—indeed, insist upon—the existence of universals?

8.11 WHY UNIVERSALS?

One reason why I admit the existence of universals, of course, is that I think that the four-category ontology offers the best (most problem-free) account of the ontological ground of the dispositional–occurrent distinction. An important part of that account is the account of laws that I have outlined above. But with a two-category ontology like Martin's, it seems, there is no scope for any account of laws other than one which represents them as consisting merely in constant conjunctions amongst particulars. However, on such a view, it seems, all such constant conjunctions are in effect nothing more than cosmic coincidences—and here I agree with Armstrong.[21] And then we are left confronting Hume's problem of induction, without very much hope of a satisfactory solution.

[21] See Armstrong's replies to Martin in Armstrong, Martin and Place, *Dispositions: A Debate*.

Someone like Martin can, of course, to some extent explain why similar macroscopic objects, at widely different times and places, behave and interact with one another in similar ways: he can explain this in terms of their having similar microscopic constitutions, that is, in terms of the similar powers and liabilities of their microstructural constituents. Ultimately, however, appeal will have to be made to the fact that similar *fundamental particles*, which lack any inner constituent structure—quarks and electrons, and so forth—have similar powers and liabilities throughout the universe in space and time. But why should this be so? Of course, it may be said that it is only because we have reason to suppose that it *is* so that we think that there are precisely these distinct 'kinds' of fundamental particle. A fundamental particle, it may be said, is wholly characterized by its distinctive powers: thus, electrons carry unit negative charge and possess a certain distinctive rest mass, whereas protons carry unit positive charge and possess another distinctive rest mass. (Forget, here, for purposes of exposition, that protons are not in fact fundamental particles, according to the current standard model of particle physics, since they are supposed to be composed of quarks.)

But recall that for someone of Martin's persuasion these *kinds* of particle are not real: that is to say, there are no such substantial universals as *electron* and *proton*. There are merely a great many individual particles with various combinations of powers, which we find to fall into certain families by virtue of the resemblances between their powers. It turns out that a great many of these individual particles have a certain combination of powers—including, say, unit negative charge and a certain distinctive rest mass—and so all of these can be denominated 'electrons'. But we still lack a satisfactory explanation of this situation, which is, *prima facie*, very remarkable and surprising. Why is it that of all the possible combinations of powers in fundamental particles, only *some* combinations are found in nature? Why, say, do we not find a particle with the rest mass of the neutron but the charge of the electron? (Again, never mind the scientific accuracy of the example: I use it purely for illustrative purposes.) It seems to me that only someone who takes laws of nature seriously, as involving universals, has an explanation for this state of affairs. We can say that there are certain laws governing the domain of fundamental particles, such as the law that electrons carry unit negative charge and the law that electrons possess a certain distinctive rest mass. Because there are these and other laws concerning *one and the same substantial kind*—the kind *electron*—there are individual objects instantiating that kind which exhibit a certain combination of powers and liabilities.[22] Individual fundamental particles exhibiting other possible combinations of powers are not found simply because there are no substantial kinds, governed by appropriate laws, for any such particles to instantiate. (I should remind the reader here that I adhere to an *immanent realist* theory of

---

[22] Here I am setting aside any questions that may legitimately be raised about whether electrons and the like really are properly to be described as 'objects': see further my *The Possibility of Metaphysics*, ch. 3.

universals, according to which there exist no uninstantiated universals and so, by my account, no uninstantiated substantial kinds.[23])

According to this view, the existence of substantial kinds and of laws governing them accounts for the truth of something like Keynes's 'principle of limited independent variety', which in turn helps us to solve Hume's problem of induction.[24] Someone who denies the reality of universals and laws, other than as arising from resemblances amongst particulars, must, it seems to me, be prepared to accept the existence of enormous and mysterious cosmic coincidences as a matter of brute, inexplicable fact. In short, then, my allegiance to an ontology including both substantial and non-substantial universals, and the laws which consist in certain of the latter characterizing certain of the former, is justified by an inference to the best explanation of the order, intelligibility and predictability of the physical universe. These are matters that I shall return to, however, in the following two chapters.

8.12 SOME COMMENTS ON MOLNAR'S THEORY OF POWERS

In an important recent book, the late George Molnar espouses a *three*-category ontology of objects, properties, and relations—all conceived as particulars, since he does not believe in the existence of universals.[25] He includes relations as comprising a distinct category because he does not consider that all relations are founded on non-relational properties of their relata—spatial relations, he thinks, provide a counterexample to any such claim. Molnar agrees with pure trope theorists like Keith Campbell that properties and relations exist, and are exclusively particulars. However, he does not agree with them that *objects*—the bearers of properties and relata of relations—are just 'bundles' of tropes, agreeing instead with C. B. Martin that properties depend for their existence and identity upon their bearers and are consequently non-transferable from one object to another. Even more importantly, he agrees with Martin in holding powers or dispositions to be real properties of objects, while rejecting Martin's view that every property is at once dispositional and qualitative in nature. Molnar thinks that there is an exhaustive and exclusive distinction to be drawn amongst properties between those that are powers and those that are non-powers, and that properties of both types exist. I shall return later to what Molnar has to say about non-powers, as I consider this part of his theory to be rather more shaky than much of the rest. Although Molnar accepts that many powers are 'grounded' in other properties and relations—most obviously, many of the powers of composite,

---

[23] See further my 'Abstraction, Properties, and Immanent Realism'.
[24] See John Maynard Keynes, *A Treatise on Probability* (London: Macmillan, 1921), p. 260. See also my 'What *is* the "Problem of Induction"?'. [25] See Molnar, *Powers*.

macroscopic objects, which he takes to be grounded in properties and relations of their parts—he holds that some powers may be, and almost certainly are, *ungrounded*, such as the powers of fundamental physical particles. He rejects, then, currently popular 'functionalist' accounts of dispositionality, which see all dispositions as being 'second-order' functional properties that are supposedly 'realized'—often, indeed, *multiply* realized—by 'first-order' categorical properties.

So what, then, *are* powers, according to Molnar? He characterizes them as having five crucial features: directedness, independence, actuality, intrinsicality, and objectivity. The directedness of powers is, for Molnar, a kind of *physical intentionality*: powers are powers *for* certain distinctive items, their possible *manifestations*, which need not exist, just as the intentional objects of mental states need not exist. The *independence* that he attributes to powers is an *ontological* independence from their manifestations: powers can exist when they are not being manifested, and even if they are never manifested. In saying that powers are *actual*, Molnar means that they are not just hypothetical in character, consisting merely in the possibility of their manifestation. In saying that they are *intrinsic* properties of their bearers, he means that they exist independently of the existence of other objects distinct from their bearers. And in saying that they are *objective* properties of their bearers, he means that they are not mind-dependent features of them. Of all these claims, the first may perhaps provoke the most hostility, particularly from philosophers sympathetic to Brentano's thesis that intentionality is the mark of the mental. Molnar makes a surprisingly good case for his position even if, in the end, one is left suspecting that at root there is nothing more than a certain analogy between the intentionality of mental states and the way in which powers are powers 'for' their manifestations. It would seem that a genuinely *intentional* mental state of any kind is always *about* certain entities—and, moreover, that a state of that kind can always be about a quite *specific* entity, rather than just about a general class of entities, however narrowly defined. On some occasion I might, for example, be thinking about *Vienna in particular*, not just about Austrian cities or European capitals in general, nor even just about cities exactly like Vienna. But a power, it seems, is never a power 'for' a quite specific manifestation, only for manifestations of a certain general type, such as bending or shattering or dissolving in water. Nor does it ever seem apt to say that a power is in any sense 'about' its possible manifestations. Molnar is fully aware of objections of this sort and addresses them thoughtfully and interestingly, although with how much success readers of Molnar's book must judge for themselves.

The case that Molnar makes for the possibility—and, indeed, the actuality—of *ungrounded* powers seems to me a very persuasive one. It involves arguing against the popular view that all dispositions require 'categorical bases'. It also involves arguing against the equally popular view that disposition statements may be analysed in terms of conditionals, or at least *entail* conditionals. C. B. Martin has already very compellingly argued against the latter view, and although David Lewis came to its defence shortly before his death, Molnar has, in my opinion, now put the matter beyond dispute. However, as I indicated earlier, I think that Molnar's discussion of the nature and existence of *non*-powers is more open to debate. Unlike a number of other contemporary metaphysicians, Molnar does not believe that *all* properties are powers, although he does not find the supposition that they are incoherent, defending it against various familiar objections. Molnar's opposition to 'pan-dispositionalism' is avowedly *a posteriori* in character, appealing to contemporary physics. But I wonder if Molnar thought this issue through carefully enough. His own view is that 'each power gets its identity from its manifestation... [and] each power has one manifestation'.[26] He means, of course, one *kind* of manifestation. So, for Molnar, each power is the very property that it is at least partly in virtue of being a power for a certain kind of manifestation. But if all properties were powers, then all *manifestations* of powers—being properties of their bearers—would themselves be powers, likewise 'getting their identity' from *their* manifestations. The problem here is not just that, on this view, 'everything is potency, and act is the mere shifting around of potencies', in the words of David Amstrong, quoted by Molnar.[27] Molnar addresses that complaint—although not, in my view, entirely convincingly. The problem, rather, is that *no property can get its identity fixed*, because each property owes its identity to another, which in turn owes *its* identity to yet another—and so on and on, in a way that, very plausibly, generates either a vicious infinite regress or a vicious circle.

But since Molnar himself is no pan-dispositionalist, let us set this issue aside. More puzzling, I think, is his view of what *do* constitute the non-powers that, he allows, actually exist. For Molnar, these are properties that are associated with certain *symmetry operations*—'S-properties'—and seem largely to be restricted to what he calls 'positional properties', such as 'spatial location, temporal location, spatial orientation, [and] temporal orientation', the only exception he mentions being 'numerical identity of parts'.[28] Oddly, it might seem, he does not regard *shape* and *size*—often cited as paradigm examples of 'categorical' properties—as non-powers. Like some other philosophers who have defended the dispositionality of geometrical properties, he points out, rightly, that such properties carry implications for the possible behaviour of objects bearing them. Thus, for example, a spherical object has, in virtue of its shape, a tendency to roll down an inclined plane—provided that, like a ball-bearing rather than a soap-bubble, it is also rigid and heavy. And a tendency to roll is, of course, a kind of disposition or power. However, it doesn't at all seem to follow that, because sphericity may *confer* such a power upon its bearer, sphericity *itself* is properly conceived to be a power, or dispositional property. If shapes themselves are powers, what are their 'manifestations'? According to Molnar, 'Shape determines the extent of contact-without-overlap that can occur between objects [and]... [t]his feature of shape is its manifestation', adding that 'this is not a contingent fact about shape'.[29] He

---

[26] Molnar, *Powers*, p. 195. [27] Ibid., p. 174. [28] Ibid., p. 160.
[29] Ibid., p. 171.

cites the example, borrowed from Martin, of the ability of a round peg—in contrast with a square one—to fit into a round hole. The peg no doubt has this ability in virtue of its shape. But I still don't see why we should have to say that its shape is, like the ability in question, a *power*. Molnar's suggestion is that the 'extent of contact-without-overlap' between the peg and any other shaped object is the 'manifestation' of the peg's roundness, conceived as a power. But I find that I can't understand this notion of 'extent of contact-without-overlap' save in geometrical terms of shape and size. For example, if one tries to fit a square peg whose sides are equal in length to the diameter d of a round hole into that hole, the minimal amount of overlap between the peg and the hole's surround is the sum of four regions of the surround each having the shape of a trilateral one side of which is an arc of a circle of diameter d and whose other two sides are straight lines of length $d/2$.

In this connection, another odd claim of Molnar's concerning shape and size is worth mentioning, namely, that he thinks that fundamental particles, such as electrons, have *neither*—that they are literally *point*-particles.[30] In opposition, I cite the following from the *Oxford Dictionary of Physics*: 'If the electron is taken to be a point charge, its self-energy is infinite and difficulties arise for the Dirac equation'.[31] The claim is important for Molnar in defending his opposition to Martin's view that all properties are *both* dispositional *and* 'qualitative' in character, because the implication seems to be that there is nothing 'qualitative' about any of the properties of fundamental particles—that all of their non-positional properties are 'pure' powers.

Let me return to the distinction, alluded to earlier but not looked upon favourably by Molnar, between *potency* and *act*—one whose roots may, of course, be traced back to Aristotle. I concur with Molnar in disliking the common practice of contrasting the dispositional with the *categorical*, partly because this invites—if, indeed, it doesn't simply arise from—the questionable assumption that what is 'dispositional' is merely *hypothetical* and so, in the idiom of possible worlds, concerns what is (categorically) the case in non-actual possible worlds. I agree with Molnar that powers reside in *this*, the actual world, and that talk of them is not to be cashed out purely in terms of what could be, or could have been, the case. I myself prefer to contrast the dispositional with the *occurrent*—for example, to constrast an object's disposition or capacity to dissolve in water with its *dissolving* in water on some occasion. An episode of dissolution in water is exactly what a 'manifestation' of the disposition of water-solubility *is*. And, I should say, such an episode precisely does *not* consist in an object's *possessing* a power, but just in its *exercising* a power: it is an *act* or *process* involving the object in question. Molnar, however, is loth to accept that 'occurrences are not dispositional'.[32] He has an ingenious argument against the view that the manifestations of powers, being

---

[30] Ibid., pp. 151–2.
[31] *The Oxford Dictionary of Physics*, 4th edn (Oxford: Oxford University Press, 2000), p. 141.
[32] Molnar, *Powers*, p. 166.

occurrences, are not dispositional. It runs as follows.[33] First, he remarks that 'objects have many accidental powers that can be acquired...during the life of the object', such as brittleness, and that '[the] event of becoming brittle is the beginning of the state of being brittle...[and] is itself a case of having the disposition'. Next, he proposes that '"dissolved" and "broken" are dispositional' (a claim which I don't intend to challenge here). Finally, he maintains that '"Dissolves" means "becomes dissolved" and "breaks" means "becomes broken"'. From all this he concludes that 'the events denoted by these expressions are themselves as dispositional as the states of the object of which they are the beginnings'. So the thought is that an occurrence of *dissolution*—that is, a manifestation of the power of solubility—is the beginning of an object's being in a dissolved state, and that since this latter state is itself dispositional in character, so too is the occurrence which is the beginning of that state.

However, a manifestation of solubility is surely *not* what Molnar here assumes it to be: it is *not* an event of an object's *beginning to be in a dissolved state*. Rather, it is a *process*—the process of dissolution—which is necessarily extended over time. Such a process essentially involves *movement* of a certain kind: for instance, in the case of the dissolution of sodium chloride in water, it involves sodium and chloride ions moving away from one another and out of their cubic lattice formation. But moving—that is, *changing location*—is most implausibly regarded as dispositional in character; nor do I think that Molnar himself can consistently maintain that it is, given that he regards spatial and temporal location as non-powers. (It may be that *instantaneous velocity* can be regarded as a power or tendency, but moving is a *process*, not something instantaneous in character.) The same point may be made regarding brittleness and its manifestation, *breaking*, which is likewise a process essentially involving movement. On Molnar's account, *becoming brittle* and *breaking* are both cases of an object's beginning to be in a dispositional state, so that the *manifestation* of brittleness is exactly on a par, ontologically speaking, with the *acquisition* of brittleness. But I submit that this betrays a misconception of the nature of the manifestation of a disposition like brittleness, which is essentially a process rather than an instantaneous event. Consequently, I see in Molnar's argument no threat to my own view that the distinction between the dispositional and the occurrent is metaphysically significant and, indeed, ontologically fundamental.

[33] Molnar, *Powers*, p. 167.

9

Kinds, Essence, and Natural Necessity

9.1 NATURAL NECESSITY AND METAPHYSICAL NECESSITY

It is commonly assumed that there is a kind of necessity, *natural* or *physical* necessity, that is in some sense 'weaker' than logical necessity and is somehow related to natural laws, or, more specifically, to causal laws—whence this supposed kind of necessity is often also called 'nomic' or 'causal' necessity. Of course, many empiricist philosophers have followed David Hume in being sceptical either about the existence or, less radically, at least about our epistemic access to any such species of necessity.[1] But in recent years the foregoing concept of natural necessity has come under attack from another quarter, namely, from certain philosophers—some of whom describe themselves as 'scientific essentialists'—who hold that natural laws are in fact necessary in the strongest possible sense: that is, who hold that the necessity of such laws is no weaker than, and just as 'absolute' as, the necessity of logical truths.[2]

Apparently, it is not possible to object to this essentialist doctrine simply on the grounds that natural laws are only discoverable empirically, because Saul Kripke has taught us that there not only may be but certainly are *a posteriori* truths that are necessary in the strongest possible sense, that is, in the sense of being true in every logically possible world.[3] The paradigm examples of such truths are truths of identity, such as the proposition that Hesperus is Phosphorus. According to Kripke, we can establish *a priori* the truth of the conditional statement 'If

---

[1] There is controversy amongst commentators as to whether Hume was an antirealist or a sceptical realist concerning causal necessity: see Galen Strawson, *The Secret Connexion: Causation, Realism, and David Hume* (Oxford: Clarendon Press, 1989). I adopt no particular position in this chapter on this historical issue.

[2] See, especially, Sydney Shoemaker, 'Causal and Metaphysical Necessity', *Pacific Philosophical Quarterly* 79 (1998), pp. 59–77, Evan Fales, 'Are Causal Laws Contingent?', in John Bacon, Keith Campbell, and Lloyd Reinhardt (eds), *Ontology, Causality and Mind: Essays in Honour of D. M. Armstrong* (Cambridge: Cambridge University Press, 1993), pp. 121–44, and Brian Ellis, 'Causal Powers and Laws of Nature', in Howard Sankey (ed.), *Causation and Laws of Nature* (Dordrecht: Kluwer, 1999). See also Sydney Shoemaker, 'Causality and Properties', in Peter van Inwagen (ed.), *Time and Cause* (Dordrecht: Reidel, 1980), pp. 109–35, and C. B. Martin, 'Power for Realists', in Bacon *et al.* (eds), *Ontology, Causality, and Mind*, pp. 175–86.

[3] The *locus classicus* is Saul A. Kripke, *Naming and Necessity* (Oxford: Blackwell, 1980).

Hesperus is identical with Phosphorus, then it is necessary that Hesperus is identical with Phosphorus'—where the necessity in question is of the strongest possible kind—even though we can only establish empirically that Hesperus is identical with Phosphorus. Given, however, that we do know, on empirical grounds, that Hesperus is identical with Phosphorus, we can then use the *a priori* truth just mentioned as the conditional premise in a *modus ponens* inference to the conclusion that, indeed, it is *necessary* that Hesperus is identical with Phosphorus. This, then, provides us with a model for the possible acquisition of *a posteriori* knowledge of truths which are necessary in the strongest possible sense. And, it is suggested, this model is applicable to our knowledge of natural laws, where these are construed as being necessary in the strongest possible sense. Let us, following the now customary usage, call this strongest possible kind of necessity *metaphysical* or *broadly logical* necessity.[4] Then, it seems, there can be no particular difficulty in maintaining that natural laws are at once metaphysically necessary and knowable *a posteriori*.

One seeming advantage in holding this view of laws is that it provides us with a perfectly straightforward account of the nature of natural necessity, for according to this view there is really only one *kind* of necessity—metaphysical or broadly logical necessity—which may be explicated, in the language of possible worlds, in terms of truth in all possible worlds. We need not, then, try to find any place for natural necessity as a kind of necessity which is somehow weaker than logical necessity but which, none the less, does not collapse into mere contingency. On the other hand, many philosophers have strong intuitions that natural laws are *not* necessary in the strongest possible sense—that a natural law which obtains in this, the actual world, need not obtain in every possible world. So we would need to be given good reasons for thinking that these intuitions are mistaken, as well as some explanation for our possession of those intuitions despite their being mistaken ones.

9.2 THE LOGICAL FORM OF LAW STATEMENTS

But why should we want to embrace any notion of natural necessity at all, whether or not this is construed as being weaker than logical necessity? One reason commonly given is that we need to embrace this notion in order to distinguish between genuine laws of nature and mere accidental regularities. (Incidentally, I should explain that here and in what follows, I distinguish between *laws* and *statements of law*, or lawlike statements—a law being the kind of state of affairs which makes a statement of law true, if indeed it is true.[5]) To use a well-known

---

[4] The term 'broadly logical necessity' is Alvin Plantinga's: see his *The Nature of Necessity* (Oxford: Clarendon Press, 1974), p. 2. See further my *The Possibility of Metaphysics: Substance, Identity, and Time* (Oxford: Clarendon Press, 1998), pp. 13 ff.

[5] Here I follow the lead of D. M. Armstrong: see his *What is a Law of Nature?* (Cambridge: Cambridge University Press, 1983), p. 8. In Chapters 11 and 12 below, I shall make it clear that and why I have doubts about regarding 'states of affairs' as truthmakers, but for present purposes I shall waive such doubts.

example, it is a merely accidental regularity that every lump of gold is less than a mile in diameter (assuming that this is indeed the case), whereas it is no merely accidental regularity that every lump of uranium is less than a mile in diameter. We want to say that no lump of uranium *could* be a mile in diameter, of natural or physical necessity, because the critical mass of uranium sufficient to produce a nuclear explosion occupies a spherical volume much less than one mile in diameter. By contrast, nothing in nature prevents us from amassing enough gold to make a lump of it one mile in diameter, even if no such lump ever has been or ever will be formed in the entire history of the universe.

At this point, we need to consider what *logical form* a natural law should be construed as having. (Strictly, it is statements of natural law that have 'logical form', but in an extended sense we can speak of the lawlike state of affairs which makes a statement of law true as having a 'logical form'.) According to the 'Humean' or 'regularity' account of laws, a law is simply a universal generalization which quantifies over particulars—in the simplest case, something of the form 'For all x, if Fx, then Gx'. Against this proposal, then, we find the objection raised that it fails to distinguish between lawlike and accidental generalizations, according both the same logical form. As part of this objection, it is observed that laws support counterfactual conditionals, whereas mere universal generalizations of the sort just described do not. Letting 'F' be 'is a lump of gold/uranium' and 'G' be 'is less than one mile in diameter', we find that 'For all x, if Fx, then Gx' does not support the counterfactual conditional 'If the universe had contained another lump of gold/uranium, it would have been less than one mile in diameter'—and yet we want to say that, in the case of uranium, the counterfactual conditional is true and should be seen to follow from the law. One proposed solution, then, is to say that laws are not mere universal generalizations but are, rather, *necessitations* of these: that is, that a law, in the simplest case, has the logical form '*Necessarily*, for all x, if Fx, then Gx', where the kind of necessity in question is 'natural' necessity, whatever that is.[6] As I shall explain later, I do not consider that this sort of proposal really provides a genuine solution to the problem in hand.

However, another sort of proposal, advanced by David Armstrong and others, is that laws do not concern particulars at all, but, rather, *universals*. According to this view, the logical form of a law, in the simplest case, is '$N(F, G)$', or 'F-ness necessitates G-ness'.[7] Such a law is then supposed to *entail* the corresponding universal generalization concerning particulars, namely, 'For all x, if Fx, then Gx', although the latter does not, of course, entail the former. One supposed advantage of this account is that it offers an explanation for the non-accidental regularity obtaining amongst particulars in terms of a special 'second-order' relation of necessitation amongst the universals exemplified by the particulars in question.

---

[6] For a partial defence of the view that laws have this form, see John Foster, 'Induction, Explanation and Natural Necessity', *Proceedings of the Aristotelian Society* 83 (1982), pp. 87–101.

[7] See Armstrong, *What is a Law of Nature?*, p. 85.

Armstrong himself regards laws as being, none the less, *contingent*, in the sense that although it may be the case that *F*-ness necessitates *G*-ness in this, the actual world, there may be other possible worlds in which *F*-ness and *G*-ness exist but in which the former does not necessitate the latter.[8] Here it is important to note that Armstrong is an 'Aristotelian' or 'immanent' realist concerning universals—that is, he holds that a universal can only exist in a world in which it is exemplified, at some time and some place, by some particular.[9] It should also be remarked that, in his more recent writings, Armstrong has tended to play down the idea that what relates the universals involved in a law is some species of 'necessitation', preferring to use causal idioms of 'bringing about' or 'making happen' to describe this alleged relation.[10] The idea, then, seems to be that particular instances of a law relating *F*-ness to *G*-ness will be particular cases of something's being *F* bringing about something's being *G*.

Finally, I should mention my own preferred view concerning the logical form of natural laws.[11] This is that, in the simplest case, a natural law has the form '*K* is *F*' (where '*K*' is a mass noun) or '*K*s are *F*' (where '*K*' is a count noun)—as, for example, 'Gold is electrically conductive', 'Electrons have unit negative charge' or 'Planets move in elliptical orbits', where these so-called *generic* propositions are by no means logically equivalent to the corresponding universal generalizations quantifying over particulars, such as 'For all x, if x is an electron, then x has unit negative charge' or 'For all x, if x is a planet, then x moves in an elliptical orbit'. Thus I too, like Armstrong, conceive of laws as involving universals rather than particulars, but distinguish, as Armstrong does not, between *substantial* universals, or *kinds*, and *non-substantial* universals, or *properties and relations*.[12] In the simplest case, a law consists in a kind's possessing, or being characterized by, a property. More complex cases involve two or more kinds standing in a relation, as with the law that water dissolves common salt or the law that positive and negative charges attract one another.

An advantage of this view over Armstrong's, I consider, is that it requires no appeal to any 'second-order' relation between universals in saying what constitutes a law. Sometimes, indeed, laws are relational in form, as in the case of the law that water dissolves common salt, but in such cases the relation involved is not a 'second-order' relation, since it is a relation in which *particular objects* can stand to

---

[8] See Armstrong, *What is a Law of Nature?*, pp. 158 ff. I should stress that I am here concerned with Armstrong's view circa 1983, not with his very latest opinion, for which see his *Truth and Truthmakers* (Cambridge: Cambridge University Press, 2004).

[9] See Armstrong, *Universals: An Opinionated Introduction* (Boulder, CO: Westview Press, 1989), pp. 75 ff.

[10] See Armstrong's replies to Smart and to Menzies in Bacon *et al.* (eds), *Ontology, Causality and Mind*, p. 172 and p. 229, and also Armstrong, *A World of States of Affairs* (Cambridge: Cambridge University Press, 1997), pp. 223 ff (he endorses the 'making happen' idiom at pp. 210–11).

[11] See further my *Kinds of Being: A Study of Individuation, Identity and the Logic of Sortal Terms* (Oxford: Blackwell, 1989), ch. 8.

[12] For Armstrong's dismissive view of kinds, see his *A World of States of Affairs*, pp. 65 ff.

one another—and in which they do so stand in particular instances of the law. A particular instance or exemplification of this law, thus, is any particular case of some water's dissolving some common salt. Another but closely related difference between Armstrong's proposal and mine is that he seems to regard all particular instances of laws as involving the exemplification of some general relation of 'bringing about' or 'making happen', whereas I favour the view that there is no such general relation, only a broad family of specific causal relations, such as dissolving, breaking, pushing, tearing, attracting, and so forth. On my view, talk of 'bringing about', 'making happen' or 'producing' is just an abstraction from more specific ways of talking about causal interaction. Moreover, my view provides a unified account both of causal laws, such as the law that water dissolves common salt, and of non-causal laws, such as the law that electrons have unit negative charge.[13]

9.3 REPLIES TO OBJECTIONS

It may be objected to my view that it incorporates no species of necessity at all into the constitution of natural laws and thus falls prey to the complaint that it fails to distinguish between lawlike and accidental generalizations. But this charge is unfair, for I do indeed distinguish between these two kinds of generalization, albeit not in terms of any notion of necessity. According to my account, a law has, in the simplest case, the form 'Ks are F' (as in 'Electrons have unit negative charge' or 'Lumps of uranium are less than one mile in diameter'), whereas accidental generalizations are mere universal quantifications over particulars, of the form 'For all x, if x is a K, then x is F', which does not entail 'Ks are F'. Indeed, on my view, 'For all x, if x is a K, then x is F', neither entails *nor is entailed by* 'Ks are F', because I hold that laws—at least, non-fundamental laws—admit of possible exceptions.[14] The law that electrons have unit negative charge is fundamental and has no possible exceptions, but a non-fundamental law, such as the law that planets move in elliptical orbits (Kepler's First Law of Planetary Motion), most certainly does have exceptions, because interfering factors (such as the gravitational influence of other planets) can prevent the instantiation of the law in particular cases.

Against me, it may perhaps be urged that, for example, it is in fact *true*, but not a law, that lumps of gold are less than one mile in diameter. But I would simply deny that this is true, even if it should happen to be true that every particular lump of gold that ever has existed or ever will exist is less than one mile in diameter. It is not true because to assert 'Lumps of gold are less than one mile in diameter', as

---

[13] Armstrong maintains that non-causal laws are supervenient: see his *A World of States of Affairs*, pp. 231ff.
[14] See further my *Kinds of Being*, ch. 8, and my 'Miracles and Laws of Nature', *Religious Studies* 23 (1987), pp. 263–78.

that sentence would standardly be understood, is to assert something false concerning the nature of the *kind* of stuff, gold, not something true about some or all particular instances of that kind that have existed or will exist. Anyone who asserted 'Lumps of gold are less than one mile in diameter', meaning thereby to say either that *some* particular lumps of gold are less than one mile in diameter or that *all* particular lumps of gold are, would be using this sentence in a non-standard and misleading way. This is clear from the fact that such a thing is *not* what is standardly meant, *mutatis mutandis*, by someone who asserts, truly, 'Lumps of uranium are less than one mile in diameter', for such a speaker evidently means thereby to say something about the nature of the *kind* of stuff, uranium.[15]

Here I should emphasize that I am by no means contending that *all* so-called generic propositions, as linguists would identify this class of propositions, are expressive of putative natural laws. For instance, 'Elephants are numerous in Africa' is not. This is because the predicate involved here does not express a property that individual elephants can possess, so that the proposition must be interpreted as expressing something about the world's elephant population—a collection or plurality of individual animals—not something about the animal kind, *elephant*, of which all of these individuals are instances. As for such propositions as 'Electrons are smaller than dogs' or 'Cats are a nuisance', which may perhaps be doubted to possess lawlike status, I would say merely that the former expresses a derivative law—one that is derivable from laws concerning the sizes of electrons and dogs respectively—and that the latter is an elliptical expression of a far-from-fundamental sociological law concerning a relationship between cats and some kinds of people. As I implied earlier, the test of whether a generic proposition is expressive of a putative law is whether it is interpretable as affirming that some *kind* is characterized by a certain *property* or, in more complex cases, that two or more kinds stand in a certain *relation*, and the last two propositions just mentioned certainly seem to pass this test. Not all putative laws are putatively fundamental laws and many of them may be of little interest to serious science.

9.4 LAWS, COUNTERFACTUALS, AND NATURAL NECESSITY

All this being granted, it may now be objected against me that, precisely because I do not distinguish between lawlike and accidental generalizations in terms of necessity, I cannot explain why laws support counterfactual conditionals. Indeed, it may be charged against me now that I have left no room for the notion of natural necessity at all and have gone over entirely to the 'Humean' camp. Not so, on

---

[15] Compare Alice Drewery, 'Laws, Regularities and Exceptions', *Ratio* 13 (2000), pp. 1–12.

either count. First of all, we need to take into account how the semantics and pragmatics of counterfactual conditionals work in these contexts, which is as follows.[16] When entertaining a counterfactual possibility, for the purposes of evaluating whether or not a given counterfactual conditional is true, we have to hold certain facts concerning actuality 'fixed', while 'unfixing' others. When evaluating a counterfactual conditional such as 'If the universe had contained another lump of uranium, it would have been less than one mile in diameter', we hold fixed, amongst other things, the laws of nature, including the law that lumps of uranium are less than one mile in diameter. This is why we can and should evaluate that counterfactual conditional as being *true*. (Obviously, we do not *always* hold fixed the laws of nature when evaluating counterfactual conditionals: for instance, we patently do not when evaluating one which begins 'If the laws of nature had been different...'. However, I am not presently concerned with such conditionals, only with conditionals such as 'If the universe had contained another lump of uranium, it would have been less than one mile in diameter', and with respect to these my remarks are, it seems clear, highly plausible.)

Here it may be objected that an adherent of the 'regularity' account of laws can make exactly the same claim and thus urge that on that account, too, the counterfactual conditional should be evaluated as being true. But in order to make this move the regularity theorist must trade upon an ambiguity. According to the regularity theorist, the law which obtains in the actual world, and which is consequently to be held fixed, consists in the fact that all *actually* existing lumps of uranium are less than one mile in diameter: but supposing that all of *these* lumps of uranium are still less than one mile in diameter in the counterfactual situation is perfectly consistent with there being an additional lump of uranium in that situation which is not less than one mile in diameter. Merely to hold fixed the existence of a true proposition of the form 'For all x, if x is a lump of uranium, then x is less than one mile in diameter' is not to hold fixed the *law*, as this is conceived by the regularity theorist. By contrast, if what is held fixed in the counterfactual situation is not merely a fact concerning all actual instances of the kind uranium but rather a fact concerning the nature of the kind uranium itself, as on my own account of laws, then indeed we can see why the counterfactual conditional should be evaluated as being true. Exactly the same may be said on behalf of Armstrong's view of laws, of course, so that this is not a respect in which my view has any advantage over his, only a respect in which both his and my views have an advantage over the regularity theorist's view.[17]

From this it will be obvious how I reply to the charge that I have left no room for the notion of natural necessity at all. My view is that there is such a thing as

---

[16] For a fuller account of my views about counterfactual conditionals, see my 'The Truth about Counterfactuals', *Philosophical Quarterly* 45 (1995), pp. 41–59.

[17] For Armstrong's deployment of a similar argument on behalf of his own view of laws, see his reply to Fales in Bacon *et al.* (eds), *Ontology, Causality and Mind*, pp. 144 ff.

natural necessity, if that is what one wants to call it, but that it is only a species of *relative* necessity. To say that a state of affairs is 'naturally necessary' is merely to say that it is a state of affairs which must be the case given the laws of nature. In possible-worlds idiom, it is something that is the case in every possible world in which the (actual) laws of nature obtain—or, at least, something that is the case in every possible world in which all such (actual) laws obtain as are *relevant* to the state of affairs in question.[18] (Thus we need not assume, perhaps implausibly, that any two laws which obtain in one world also co-obtain in any other world in which either of them obtains.) Analogously, something is 'legally' necessary, or obligatory, in the forensic sense, if it must be the case given the laws of the land—as, for example, it is legally necessary, or obligatory, that a citizen does not steal or commit murder. Against this view, the scientific essentialist holds, as we saw earlier, that natural necessity is in fact nothing other than metaphysical necessity, and so is 'absolute' necessity. The scientific essentialist may agree that a state of affairs is naturally necessary if and only if it is the case in every possible world in which the (actual) laws of nature obtain—but since he also contends that these laws are themselves metaphysically necessary and so obtain in every possible world whatever, this is consistent with his view that a naturally necessary state of affairs is one which obtains in every possible world whatever. We have yet to see in detail what might be said for or against this essentialist doctrine. But all I want to claim at this stage is that there is an alternative and ostensibly coherent account of laws and natural necessity which does not treat laws as mere regularities amongst particulars, which treats such laws as metaphysically contingent, and which none the less accommodates both an explicable notion of natural necessity and our conviction that laws of nature support counterfactual conditionals.

Indeed, there is more than one such account of laws, since both Armstrong's account and my own meet these criteria. But as for the other 'non-Humean' account of laws mentioned above, according to which laws are *necessitations* of universal quantifications over particulars, my opinion is that this does not meet the criteria in question, because it leaves unexplained the nature of the necessity at issue. Plainly, one cannot non-circularly explain this notion of necessity in terms of truth in all worlds in which the actual laws of nature obtain, if one appeals to the very notion of necessity in question in explaining what constitutes a law. Note here that there is no such circularity in Armstrong's (original) view, according to which a law has, in the simplest case, the form '*F*-ness necessitates *G*-ness'. For it is clear that the necessitation *relation* between universals which, on this view, helps to constitute a law of nature is not to be confused with the 'natural necessity' which may be said to characterize states of affairs, where this is explicated in terms of the obtaining of those states of affairs in every possible world in which the

---

[18] But see further my 'Miracles and Laws of Nature' for reasons to modify this proposal in certain ways which, however, do not undermine the argument of the present chapter. For more on the distinction between 'relative' and 'absolute' necessity, see my *The Possibility of Metaphysics*, pp. 18 ff.

(actual) laws of nature obtain.[19] On the other hand, this very fact suggests that it is at least misleading to speak of that alleged relation as one of 'necessitation' and this may help to explain Armstrong's tendency to downplay such talk in his more recent writings.

9.5 SCIENTIFIC ESSENTIALISM AND THE IDENTITY CONDITIONS OF PROPERTIES

Let us assume, in what follows, that laws of nature do indeed concern universals rather than particulars. Scientific essentialists may be expected to agree, because one of the primary claims often made in support of their view is that natural kinds and properties depend for their very identity upon the laws into which they enter.[20] Thus, one contention is that a property owes its identity to the contribution that it makes to the (perhaps conditional) causal powers of physical objects possessing that property, as determined by the causal laws governing interactions involving such objects. For example, if we ask what property *mass* is, we may be told that mass is that property in a body which, amongst other things, makes it necessary for a force to be applied to the body in order to accelerate it, in accordance with Newton's Second Law of Motion, $F = Ma$. Again, we may be told that *sphericity* is that property which, amongst other things, makes a body which possesses it liable to roll down an inclined plane (provided that the body is also rigid and subject to a gravitational force), in accordance with various natural laws. In support of this view, it may be pointed out that we can only *detect* the properties of physical objects by interacting with them or with other objects affected by their activities. So it seems reasonable to adopt something like the following criterion of identity for (physical) properties: the property of being F is identical with the property of being G if and only if F and G make the same contribution to the causal powers of physical objects possessing them. Then, given that the causal powers of objects are determined by the natural laws governing the properties (universals) which those objects possess, the foregoing criterion seems to reduce to this: the property of being F is identical with the property of being G if and only if F and G enter into the same laws in the same ways.

Thus, consider all the laws into which the property of mass, M, enters, such as the law that force equals the product of mass and acceleration ($F = Ma$) and the law (Newton's Law of Gravitation) that the gravitational force between two bodies is proportional to the product of their masses and inversely proportional to the

---

[19] Of course, Armstrong himself is not a realist concerning possible worlds: see his *A Combinatorial Theory of Possibility* (Cambridge: Cambridge University Press, 1989) and his *A World of States of Affairs*, pp. 172 ff. For my own view of the ontological status of possible worlds, see my *The Possibility of Metaphysics*, pp. 256 ff. Nothing that I say in the present chapter in the language of possible worlds is intended to imply a commitment to their reality.

[20] See the papers cited in note 2 above.

square of the distance between them ($F = GM_1M_2/r^2$). (Purely for the purposes of argument, I am assuming the truth of classical Newtonian physics rather than of Einsteinian relativistic physics.) Then, any property, P, which entered into exactly the same lawful relations with other properties as does M, the property of mass, would have to be regarded as being *identical* with the property of mass, since there would be no possible way of distinguishing a body's possession of P from its possession of M in terms of any effect it could have on us or other bodies. But then, it may seem, we are committed to the essentialist doctrine that the laws of nature are *metaphysically* necessary—that any law which obtains in this, the actual world, must obtain in every possible world. For if the law involves certain properties (universals), P_1 to P_n, and these properties, like all properties, owe their very identity to the laws in which they are involved, then any world in which those very properties exist will be a world in which the law in question obtains, and any world in which they do not exist will be a world in which the law obtains vacuously.[21] Hence the law will obtain in any world whatever and so be metaphysically necessary.

Unfortunately for the scientific essentialist, this reasoning is quite fallacious, because it illicitly assumes that a criterion of identity for properties in this, the actual, world must also serve as a principle of so-called *transworld identity*. Consider, by way of analogy, Donald Davidson's famous (though now abandoned) criterion of identity for events, whereby event e_1 is identical with event e_2 if and only if e_1 and e_2 have exactly the same causes and effects.[22] However reasonable this criterion may be—and, certainly, it is not easy to dispute its *truth*, even if it may be objected that it is implicitly circular as a criterion of identity[23]—it cannot at all plausibly be made the basis of a principle of transworld identity for events, because we have a strong intuition that any particular event *could* have had at least some causes and effects other than those that it actually has. Indeed, if we do turn this criterion into a principle of transworld identity, we saddle ourselves with the doctrine that all singular causal relations are metaphysically necessary: that if an event e_1 in fact caused an event e_2, then e_1 causes e_2 in every possible world in which either of these events exists.

Now, perhaps the scientific essentialist would be happy to accept this too—indeed, it may be that, to be consistent, he must accept it. But this does not detract from my point that, so far, we have been given no good reason to adopt this position, nor the corresponding position regarding the transworld identity of properties. Quite generally, one cannot advance directly from an *intra*world criterion of identity to an *inter*world criterion of identity. As a *reductio ad absurdum* of the supposition

---

[21] See further Ellis, 'Causal Powers and Laws of Nature', p. 30.

[22] See Donald Davidson, 'The Individuation of Events', in his *Essays on Actions and Events* (Oxford: Clarendon Press, 1980), pp. 163–80. He abandons the proposal in his 'Reply to Quine on Events', in Ernest LePore and Brian McLaughlin (eds), *Actions and Events: Perspectives on the Philosophy of Donald Davidson* (Oxford: Blackwell, 1985), pp. 172–6.

[23] For more on the circularity objection to Davidson's criterion, see my *The Possibility of Metaphysics*, p. 43.

that one can, consider the claim that because two material objects of the same kind cannot occupy the same place at the same time—so that spatiotemporal location provides a criterion of identity for material objects of any given kind—it therefore follows that no material object of a given kind could have occupied a spacetime location different from its actual spacetime location—that this very chair, say, could not have been located where that chair is now. Plainly, no such conclusion does follow.

At this point, the scientific essentialist may reply that properties are relevantly different from particular objects and events where these issues of identity are concerned.[24] Objects and events are non-repeatable entities with unique spatiotemporal locations, so that we can identify them in this world in ways which differ from the ways in which we identify them 'across' worlds. He may try to press home the point by urging that it simply doesn't make sense to suppose that the very same property—mass, let us say—might, in another possible world, make the same contribution to the laws of that world that, let us say, electrical charge does in this, the actual world. Or, to take another example, he may urge that it simply doesn't make sense to suppose that there might be a world in which the kind *electron* exists, but in which the law is that electrons have unit *positive* charge instead of unit negative charge. These otherworldly 'electrons' would surely just be *positrons*.

However, one can oppose scientific essentialism without being driven to accept these extreme consequences. If we consider less extreme variations in the laws, the scientific essentialist's case seems much less plausible. Must we say, for instance, that a world in which the law of 'gravitation' is an inverse cube law, or a law making the force inversely proportional to a non-integral power of the distance very slightly different from 2, is a world in which the property of *mass* does not exist? What about a world in which the universal constant of gravitation, G, has a slightly different value from its value in this world—is that likewise a world in which the actual world's property of mass does not exist? Again, must we say that a world in which the ratio between the value of G and the value of the charge on the electron differs somewhat from the ratio obtaining in the actual world is a world in which neither mass as we know it nor electrons as we know them exist? It would be highly speculative to respond to any of these questions by urging that a future physics will one day discover mathematically necessary connections between all the constants of nature which render it metaphysically impossible that these constants should have values at all different from their actual values. Certainly, nothing in current physics can give us any confidence in such a prospect. Indeed, we cannot even treat as sacrosanct the idea that the so-called 'constants' of nature, such as G, really are *constant over time*.[25] But if G were discovered to be gradually

---

[24] One essentialist who does at least discuss the matter—though not, to my mind, entirely satisfactorily—is Sydney Shoemaker: see his 'Causal and Metaphysical Necessity'.

[25] In fact, E. A. Milne's cosmological theory of kinematic relativity, which was perfectly respectable from a scientific point of view, proposed a secular increase in the value of the 'constant' of gravitation: see Hermann Bondi, *Cosmology*, 2nd edn (Cambridge: Cambridge University Press, 1961), ch. 11.

increasing over time, would it not be absurd to conclude that the property of mass which exists now is not the same property of 'mass' which existed at some time in the past?

Let it be clear that I do not want to challenge scientific essentialism in its entirety. I am prepared to accept, or at least to countenance, the Kripke/Putnam view that water is essentially H_2O and that common salt is essentially sodium chloride, NaCl.[26] What I dispute is that it in any way follows from this that the *natural law* that water dissolves common salt is metaphysically necessary. In every possible world in which both water and common salt exist, I am prepared to accept, those substances are composed of molecules consisting, respectively, of two hydrogen ions and one oxygen ion and of one sodium ion and one chlorine ion. (Though, more accurately, we should say that sodium chloride in its crystalline state consists in a cubic lattice structure held together by electrostatic forces between its constituent sodium and chlorine ions, whereas the hydrogen and oxygen in water really do exist in the form of discrete H_2O molecules.) Likewise, I am prepared to accept, for instance, that an oxygen atom, in any possible world in which it exists, consists of a nucleus containing eight protons and a closely comparable number of neutrons, surrounded by eight orbital electrons. Now, it is true that the chemical interactions of water and common salt are determined by their atomic structure and the laws governing their atomic constituents, especially their orbital electrons. But, provided that we can accept that atomic constituents of those very kinds—electrons, protons and so forth—can exist in worlds in which the constants of nature have somewhat different values, or in which power laws differ somewhat in the values of the exponents involved, then we may have to accept that water and common salt can exist in worlds in which the characteristic chemical interactions of those substances differ significantly from those which they exhibit in this, the actual world—perhaps even to the extent of there being worlds in which, although water and common salt both exist, the former lacks the power to dissolve the latter, so that it is not a *law* in that world that water dissolves common salt. (I shall return to the issue in the next chapter.)

9.6 HOW IS KNOWLEDGE OF LAWS POSSIBLE?

At this point I should like to challenge one of the mainstays of scientific essentialism, namely, its presumption that the Kripkean explanation of how *a posteriori* knowledge of metaphysically necessary truths is possible is extensible to our knowledge of natural laws, conceived as constituting metaphysically necessary truths. We saw earlier how this explanation works with a simple truth of identity concerning particular material objects, such as the fact that Hesperus is identical with

---

[26] See Kripke, *Naming and Necessity*, and Hilary Putnam, 'The Meaning of "Meaning"', in his *Mind, Language and Reality: Philosophical Papers, Volume 2* (Cambridge: Cambridge University Press, 1975), pp. 215–71.

Phosphorus. We know *a priori* that if Hesperus is identical with Phosphorus, then it is metaphysically necessary that Hesperus is identical with Phosphorus (accepting, for current purposes, the validity of the Barcan–Kripke proof of the necessity of identity).[27] And we can establish empirically that Hesperus *is* identical with Phosphorus, because we can determine by astronomical observation that their orbital positions coincide. Hence we can infer, by *modus ponens*, that it is metaphysically necessary that Hesperus is identical with Phosphorus. But, of course, we can only establish the empirical truth that Hesperus is identical with Phosphorus because we can appeal in this case to another *a priori* principle, namely, a criterion of identity for *planets*, conceived as a kind of material object. This criterion tells us that no two planets can exist in the same place at the same time, whence we can conclude, from the astronomical observation that Hesperus and Phosphorus coincide in their orbital positions, that Hesperus is identical with Phosphorus.

However, as will be obvious, this criterion of identity for planets is an *intra*world criterion, not an *inter*world or 'transworld' criterion. And that fact is vital to our ability to settle the identity question concerning Hesperus and Phosphorus empirically. If the only admissible principle for determining the identity or diversity of Hesperus and Phosphorus were a *transworld* principle, then we would not, after all, be able to determine their identity empirically, for it would not then suffice to adduce facts concerning those planets which obtain in the actual world and which are therefore empirically accessible to us, such as facts concerning their actual orbital positions—we should need also to adduce facts concerning them which obtain in every other possible world in which they exist, and such facts are not empirically accessible to us.

Now we are in a position to construct a dilemma for the scientific essentialist, as follows. Either the essentialist allows, in the case of properties and kinds, that a distinction is to be drawn between an intraworld and an interworld criterion of identity for such entities, or else he does not allow this and holds that the same criterion necessarily governs both intraworld and interworld identity in such cases. If the essentialist adopts the first option, then his claim that laws of nature are metaphysically necessary is undermined. For according to this option, while it may be accepted, say, that the property of being F is identical with the property of being G if and only if F and G enter into the same (actual) laws in the same ways, it must at the same time be agreed that, because this is only an *intra*world criterion of identity, it is also metaphysically possible for F or G to have entered into different laws.[28] On the other hand, if the essentialist adopts the second option, as he really

---

[27] In point of fact, I have doubts about the Barcan–Kripke proof: see my 'On the Alleged Necessity of True Identity Statements', *Mind* 91 (1982), pp. 579–84.

[28] I do not mean to suggest that I endorse this intraworld criterion of identity for properties myself: in fact, I reject it on the grounds that it is circular. For my own account of the identity conditions of properties (universals), see my 'Abstraction, Properties, and Immanent Realism', in Tom Rockmore (ed.), *Proceedings of the 20th World Congress of Philosophy, Volume 2: Metaphysics* (Bowling Green, OH: Philosophy Documentation Center, 1999), pp. 195–205.

seems bound to do, he must relinquish the claim to be able to apply the Kripkean model of how our *a posteriori* knowledge of natural laws is possible and hence render altogether mysterious how empirical knowledge of such laws, conceived as metaphysically necessary, is available.

Consider, for instance, the law that electrons have unit negative charge. How do we know that this law is true? The obvious thing to say is that we know that it is true, or at least have very good empirical grounds for believing that it is true, because scientists have conducted many experiments (such as Millikan's oil drop experiment) to measure the charge on particular electrons and have found that it consistently takes a certain negative value, which has consequently been defined as the unit negative charge. But to conduct such an experiment we must in principle be able to identify some particular entity empirically as being of the kind *electron* and be able to determine by empirical means that this entity possesses the property *unit negative charge*. Yet how can we determine empirically, in any given case, that we are indeed confronted with exemplifications of the universals *electron* and *unit negative charge*, if the very identity of these universals turns on what laws they enter into in every possible world (including, of course, the actual world)? If it is part of the very essence of electronhood and of unit negative charge that electrons have unit negative charge, because this law involving them obtains in every possible world, how do we know that our world is one in which these very universals do indeed exist and are exemplified by particular entities, as opposed to being a world in which the law in question is merely vacuously true and the entities that we experiment upon exemplify quite different universals?

The scientific essentialist may be tempted at this point to appeal to something like Hilary Putnam's notion of a 'stereotype'.[29] According to Putnam, a natural kind term, such as 'water' or 'lemon', has associated with it a bundle of stereotypical properties—in the case of 'lemon', the properties, say, of being yellow, juicy and acidic—which do not necessarily belong to every individual exemplar of the natural kind in question but which do belong to 'typical' exemplars and which can therefore be used, albeit only defeasibly, to identify something as being an exemplar of that kind. However, in the first place, if these stereotypical properties are indeed genuine *properties*—universals—possessed by individual objects, then the scientific essentialist ought, in all consistency, to say that these properties too owe their very identity to the laws into which they enter, in which case it becomes equally mysterious how we can ever know empirically that anything exemplifies *them*. Secondly, when we are concerned with fundamental kinds such as the kind *electron*, it appears that we can no longer suppose that there are stereotypical properties associated with the kind but which not every individual exemplar of the kind necessarily possesses: all electrons, it seems, are necessarily exactly alike in respect of their intrinsic properties. Hence, the proposed model for our ability to identify exemplars of natural kinds empirically, even if it served the scientific essentialist's

---

[29] See Putnam, 'The Meaning of "Meaning" '.

purposes in some cases—which I do not think it does—would not serve them in the case of fundamental natural kinds. I conclude that the scientific essentialist has an undischarged burden of explaining how, in general, our empirical knowledge of natural laws can be possible if, as he maintains, the very essence of a property resides in its lawlike connections with other properties.

This, however, is not my final word on the matter. My aim here has been to challenge the scientific essentialist's case for supposing that properties *quite generally* have their 'transworld' identities fixed by the laws in which they figure. I certainly do think that this case has not been made out adequately and that the supposition in question has the consequence of rendering our knowledge of laws problematic. But, as will emerge in the next chapter, I think that there are other and better reasons for thinking that *some* laws, including—*when it is properly interpreted*—the law that electrons possess unit negative charge, are indeed metaphysically necessary.

10

Categorial Ontology and Scientific Essentialism

10.1 LAWS AND NECESSITY REVISITED

As I remarked in the previous chapter, it is often supposed that there is a species of necessity—*natural, physical,* or *nomic* necessity—which is somehow weaker than logical or metaphysical necessity and is somehow grounded in or flows from the laws of nature. Invoking the language of possible worlds, thus, it is sometimes said that an event or state of affairs is naturally necessary just in case it occurs or obtains in every possible world in which the laws of nature hold. Here, of course, it is being assumed that by 'the laws of nature' is meant the natural laws which hold in this, the actual world. It is conceded, thus, that in some metaphysically possible worlds other laws may hold, in which events may occur or states of affairs obtain which are, relative to the actual world, naturally impossible. This, indeed, is precisely why natural necessity is thought to be 'weaker' than logical or metaphysical necessity. In effect—as we saw earlier—this is to conceive of natural necessity as a sort of *relative* necessity, in contrast with *absolute* necessity of the logical or metaphysical variety.[1] On this view, the naturally necessary is that which is necessary relative to, or given, a certain complete system of laws, where by a 'complete' system of laws I mean a set of laws which together constitute *all* the laws obtaining in some possible world. That which is *actually* naturally necessary is, then, that which is necessary relative to the actual system of laws. In other words, something is *naturally* necessary, on this view, just in case it is *metaphysically* impossible that the actual laws should hold and this thing not be the case.

But what, really, *is* a natural law? To approach this question, we can do no better, perhaps, than to cite some paradigm examples, such as Newton's laws of motion and of gravitation. Of course, Newtonian physics has now been superseded by relativistic physics and quantum mechanics, but let us ignore this for present purposes. After all, the view we are now considering maintains that

---

[1] For more on this distinction, see Bob Hale, 'Absolute Necessities', in James E. Tomberlin (ed.), *Philosophical Perspectives, 10: Metaphysics* (Oxford: Blackwell, 1996), and my *The Possibility of Metaphysics: Substance, Identity, and Time* (Oxford: Clarendon Press, 1998), pp. 13 ff.

other laws than the actual laws could have held—*do* hold in some metaphysically possible worlds—so let us pretend that our world, the actual world, is one such world, as indeed it was at one time thought to be. What exactly is Newton's law of gravitation? It is easy enough to state it: the gravitational force of attraction between two massive bodies is proportional to the product of their masses and inversely proportional to the square of the distance between them. In symbolic form: $F = GM_1M_2/r^2$, where G is the universal constant of gravitation. But we must distinguish between these statements or formulations of the law, which are mere linguistic or symbolic representations, and the state of affairs that they represent as obtaining. In the language of truthmaking, we must distinguish between a law statement and its truthmaker. Many philosophers mean by a 'law' a true law statement—and this practice is unobjectionable in itself, so long as it is recognized that something must actually exist to make such a statement true, if indeed it is true. I prefer to follow David Armstrong's practice and call the *truthmaker* of a law statement a law, rather than the law statement itself.[2] The law statement itself we can simply call a law statement.

However, just knowing how to state laws and that they must have truthmakers doesn't tell us what the truthmakers of law statements are—doesn't tell us, that is, what *laws* are. We mustn't fall prey to the lazy assumption that we can read off the ontological character of laws from the syntax of law statements. Laws are often stated in the form of so-called universal generalizations. In the case of Newton's law of gravitation, thus, such a statement of the law would be 'Every pair of massive bodies is subject to a mutually attractive force proportional to the product of their masses and inversely proportional to the square of the distance between them'. Notoriously, however, this way of stating laws fails to make salient the difference between lawlike and so-called 'accidental' generalizations, as we remarked in Chapter 9 above. The distinction is, once again, well illustrated by the following pair of generalizations: 'Every sphere of solid gold is less than one mile in diameter' and 'Every sphere of solid uranium is less than one mile in diameter'. Although each generalization is very probably true, the first is plausibly only accidentally true, because it so happens that that much gold has never been and never will be gathered together. The second is true for a different reason, namely, because it is naturally impossible for so much uranium to be gathered together, since the mass of such a sphere would far exceed the critical mass at which uranium explodes.

The very way in which I have expressed this explanation suggests another and more perspicuous way of stating laws. Consider the statement: 'Uranium explodes when it exceeds its critical mass'. This seems to express a *law*, no less, concerning uranium. It states a fact concerning the nature of this element, just as 'Gold

---

[2] See David M. Armstrong, *A World of States of Affairs* (Cambridge: Cambridge University Press, 1997), p. 2. As I have made clear in previous chapters, I don't mean to commit myself here to Armstrong's view that states of affairs are the *basic* truthmakers of all truths: see further Chapters 11 and 12 below.

dissolves in aqua regia' states a fact about the nature of gold and thereby expresses a law. I venture to affirm, indeed, that many and perhaps even all natural laws just *are* facts about the natures of certain natural kinds of things or stuff. Even Newton's law of gravitation can be understood in this way, as a fact about the nature of massive bodies quite generally. It is true that many very different kinds of things can be massive, that is, possess mass—for instance, stars, trees and fish. However, what all of these things have in common is that they are composed of matter: each of them is a massive thing because each of them is constituted, at any time at which it exists, by a certain mass of matter, albeit by different such masses at different times. Strictly speaking, then, Newton's law of gravitation is a law concerning the nature of masses of matter. The law is that masses of matter attract each other with a degree of force which is proportional to the product of the quantities of their mass and inversely proportional to the square of the distance between their centres of mass.

10.2 NATURAL LAWS AND NATURAL KINDS

I have just suggested that natural laws are facts about the natures of natural kinds of things or stuff. But what is a natural kind and what should we understand by the 'nature' of such kind? It seems clear that what we must be talking about here are items which belong to the category of *universals* rather than to that of *particulars*. Of course, there are many philosophers who are sceptical about the existence of universals and even those who accept their existence do not all agree about how the distinction between universals and particulars is properly to be drawn, nor about the manner in which universals exist—for instance, whether they are 'immanent' or 'transcendent' and (if this is a distinct issue) whether or not they exist in space and time. Let me, however, briefly reiterate my own stance on these matters. I believe in the existence of universals. I believe that what distinguishes universals from particulars is that universals can be instantiated—they can have *instances*—whereas particulars cannot. Indeed, I believe that every particular just *is* an instance of some, and maybe of more than one, universal. I believe that universals are immanent, at least in the sense that no universal can fail to be instantiated. At the same time, I hold that universals are not spatiotemporally located entities: they do not literally exist *in* the places and *at* the times in and at which their particular instances exist. Were they to do so, they would have to be capable of *multiple location*, given that most universals have many different particular instances, which themselves exist in and at different places and times. But I cannot understand how anything could literally be multiply located in the way in which universals would have to be if they were to exist in and at the places and times of all of their particular instances.

Now I need to recall a distinction that I draw between two importantly different categories of universals. On the one hand, we have *kinds*, whose particular instances are *objects*. On the other hand, we have *properties and relations*, whose particular instances are items for which no universally accepted name is available, being variously called tropes, modes, moments, individual accidents, particularized properties and relations, or property- and relation-instances. I favour the term *mode*, distinguishing between relational and non-relational modes. Also, to avoid any possible ambiguity, I shall use the term *attribute* to refer to properties and relations conceived as universals. Thus we arrive at my *four-category ontology* embracing two categories of universals—kinds and attributes—and two corresponding categories of particulars—objects and modes. In this system, it is important not to confuse the relationship between particulars and universals, which is instantiation, with the relationship between kinds and attributes or between objects and modes—which I call *characterization*. An object is characterized by its modes—its particular shape, mass, charge, or colour, for instance. Similarly, a kind is characterized by its attributes: for instance, gold is characterized by its attributes of being yellow, shiny, heavy, malleable, ductile, and soluble in aqua regia.

Let us focus for a moment on one characteristic of gold, its solubility in aqua regia. Earlier I remarked that 'Gold dissolves in aqua regia' is a law statement, and indeed a true one. What makes it true, then, is a law of nature—a law concerning the nature of gold as a natural kind of stuff. As I see it, the law simply consists in the fact that this kind—a universal—is characterized by solubility in aqua regia—another universal, although this time an attribute rather than a kind. More precisely, the law consists in the fact that two different kinds—gold and aqua regia—are characterized by a dyadic relation which holds between them: the relation in which two kinds of stuff stand when one of them is soluble in the other. Call this relation the *dissolution* relation. Then the law in question consists in the holding of the dissolution relation between aqua regia and gold. One commonplace way of stating the law expresses this fact particularly explicitly, namely, 'Aqua regia dissolves gold'. Here we have a statement of the law which, in my view, accurately represents its ontological structure, with the common nouns 'aqua regia' and 'gold' denoting two natural kinds of stuff and the transitive verb 'dissolves' denoting an asymmetrical relation in which those kinds stand to one another.[3] Moreover, since the process of one thing's dissolving another is clearly a causal one, we can and should say that the law in question is a *causal* law.

10.3 CAUSAL POWERS AND LIABILITIES

If I am right about all this, then laws exhibit, at the level of universals, ontological structures which exactly parallel the ontological structures of facts concerning the

---

[3] I say more about and in defence of this conception of laws in my *Kinds of Being: A Study of Individuation, Identity and the Logic of Sortal Terms* (Oxford: Blackwell, 1989), chs. 8–10.

particulars which instantiate those laws. A particular instance of the law that aqua regia dissolves gold is a particular case of some aqua regia's dissolving some gold—the sort of fact that is reported by a singular statement of the form 'This aqua regia is dissolving this gold'. Such a fact consists in the holding of a *particular* relation—a relational mode, as I call it—between two particular objects, one of them an instance of the kind aqua regia and the other an instance of the kind gold. The relational mode is, in turn, an instance of the relational universal which we earlier elected to call the dissolution relation and which is itself an ingredient in the law that aqua regia dissolves gold. Now, of course, just because it is a law that aqua regia dissolves gold, it doesn't follow that every particular instance of aqua regia is dissolving some instance of gold. In order for some aqua regia to dissolve some gold, it is at least necessary that they be in contact with one another. Even so, we can truly say of any particular instance of aqua regia that it has the causal *power* to dissolve gold and of every instance of gold that it has the causal *liability* to be dissolved by aqua regia. Indeed, in my view, the fact that a particular instance of aqua regia has the power to dissolve gold is simply a consequence of two more fundamental facts: the fact that it is an instance of aqua regia and the fact—the law—that aqua regia dissolves gold. In short, particular objects derive their powers and liabilities from the laws governing the kinds which they instantiate.

This view of powers and liabilities will, of course, be rejected by philosophers who deny the existence of universals. Many of them will no doubt urge that my account of the relationship between the powers and liabilities of particular objects and natural laws is precisely back to front. They will say that certain laws of nature obtain—such as the law that aqua regia dissolves gold—because similar particulars possess similar powers and liabilities: because, for instance, every particular quantity of aqua regia has a power to dissolve gold and every particular body of gold has a liability to be dissolved by aqua regia. For these philosophers, it would seem, laws must simply consist in regularities or uniformities concerning the powers and liabilities of particular objects. But how are such uniformities to be explained? Without the possibility of any appeal to universals, it might seem that such uniformities can amount to no more than cosmic coincidences. *I* can explain the uniform possession of a power to dissolve gold by all particular bodies of aqua regia by the facts that all of these particulars are instances of the same kind of stuff and that this kind of stuff is one in whose nature it is to dissolve gold—the latter fact constituting a natural law governing the kinds aqua regia and gold. But what can the opponent of universals say?

Let us call the opponent of universals the *particularist*, this being more perspicuous than the usual term 'nominalist'. The particularist may say that the reason why similar objects possess similar powers and liabilities is that these powers and liabilities are grounded in similar structural or 'categorical' features of those objects. For instance, it may be said that every sodium chloride crystal's liability to be dissolved by water is explained by the fact that the liability is, in each case, grounded in the cubic lattice structure of the crystal, in which sodium and chloride ions are held together by the electrostatic forces between them, plus the fact that every water molecule has a dipole moment which can disrupt those cohesive forces. However, such an explanation itself appeals to the powers and liabilities of certain kinds of ions and molecules, so we can again ask why it is that, for example, all sodium ions have similar powers and liabilities. Here it may again be contended that this is so because those powers and liabilities are grounded in structural similarities, but this time at the subatomic level, involving the orbital electrons of sodium ions. But, once more, any such explanation must appeal to the powers and liabilities of further entities, namely, subatomic particles, such as electrons—for instance, to such powers and liabilities as their electrical charge and inertial rest mass. However, electrons are thought to be *fundamental* particles, with no internal structure and no constituents, so that their powers and liabilities must, it seems, be ungrounded or basic. Hence, one cannot explain why all electrons have similar powers and liabilities in terms of structural or 'categorical' similarities between them. Above all, one cannot explain in these terms why all electrons have the same *combination* of powers and liabilities, and why certain other combinations of powers and liabilities are not to be found in any actually occurring species of fundamental particle.

The particularist, it seems, must simply accept these as brute facts, which on his account look to be accidents or coincidences of cosmic proportions. What *I* can say, however, is that the uniformities in question are explained by the fact that electrons are all particular instances of the same fundamental natural kind, which is governed by a number of laws linking this kind with certain attributes. And I can explain the absence of other regularly occurring combinations of powers and liabilities in terms of the non-existence of any *kinds* of particle governed by suitable laws. It is true that these facts concerning kinds seem to be 'brute' facts, but at least we have reduced the number of brute facts to a few and are not left with inexplicable cosmic coincidences in the realm of particulars.

10.4 COMPARISONS WITH OTHER VIEWS

Having explained why I reject particularism, I want now to compare my position with those of two other opponents of particularism—David Armstrong and Brian Ellis.[4] Armstrong favours a two-category ontology of particulars and universals. On the side of particulars, he recognizes the existence of particular objects, but not the existence of particular properties and relations—that is, tropes or modes. On the side of universals, he recognizes the existence of properties and relations—that is, attributes (as I call them)—but not the existence of kinds (not, at least, as universals that are irreducibly distinct from attributes or combinations of

---

[4] See, especially, Armstrong, *A World of States of Affairs*, and Brian Ellis, *Scientific Essentialism* (Cambridge: Cambridge University Press, 2001).

attributes). Consequently, for Armstrong, no distinction can be drawn between instantiation and characterization: for him, particulars can only be related to properties as instantiating or exemplifying them. Now, like me, Armstrong holds that laws involve universals, but unlike me he says that a law consists in the holding of a 'second-order' relation between 'first-order' universals—a relation which, at one time, he spoke of as 'necessitation'. Thus, in the simplest sort of case, an Armstrongian law is supposed to consist in one property's necessitating another. So, for instance, it might be a law that *F*ness necessitates *G*ness, which can be written symbolically as '$N(F, G)$'. This in turn is supposed to have as a consequence that any particular object which exemplifies *F*ness also exemplifies *G*ness. In other words, '$N(F, G)$' is supposed to entail '$(\forall x)(Fx \to Gx)$'.[5]

It is in this way that Armstrong endeavours to explain the existence of lawlike uniformities amongst particulars, such as the fact that all bodies orbiting stars move in elliptical paths. In this case, the lawfully connected universals would be the property of orbiting a star and the property of moving in an elliptical path. Like me, then, Armstrong is not faced with having to say that such a uniformity is, ultimately, just a cosmic coincidence, but can explain it as issuing from a law connecting certain universals. However, because Armstrong does not recognize the distinction between kinds and attributes and therefore cannot speak, as I do, of attributes *characterizing* kinds, he needs, as we have just seen, to invoke a second-order relation to connect the universals involved in a law, the relation of 'necessitation'. Since this supposed relation has seemed mysterious to Armstrong's critics, my account of laws has a distinct advantage over his in this respect. However, as we shall see, matters are complicated somewhat by more recent developments in Armstrong's thinking to which we shall return shortly.

Where Armstrong's ontology is sparser than mine, Ellis's is more abundant. He operates with a six-category ontology.[6] In addition to the four fundamental categories that I acknowledge, he includes two more. On the side of universals, he includes also the category of kinds of events and processes; and on the side of particulars, he includes particular instances of those kinds, that is, particular events and processes. My reason for ignoring these categories is not that I do not believe in the existence of events and processes, but just that I do not think that they or the kinds to which they belong are ontologically fundamental entities. Mine is a substance ontology in the Aristotelian tradition and I accordingly consider that events and processes just are, ultimately, changes or sequences of changes in the properties and relations of particular objects.[7] To be fair to Ellis, he seems to acknowledge that such a reduction may be available, but prefers to keep an open mind about this rather than to commit himself definitely either way. Ellis, like Armstrong and myself, holds that laws involve universals and serve to explain

---

[5] See David M. Armstrong, *What is a Law of Nature?* (Cambridge: Cambridge University Press, 1983), Part 2.   [6] See Ellis, *Scientific Essentialism*, p. 74.
[7] See further my *A Survey of Metaphysics*, ch. 13.

uniformities amongst particulars rather than just consisting in such uniformities. But there is another respect in which Ellis departs from what both Armstrong and I say about laws—although, as we shall see, Armstrong seems lately to have been shifting his opinion on this matter. This is that Ellis holds laws to be metaphysically necessary, whereas Armstrong and I hold them to be metaphysically contingent (or, at least, Armstrong did so until very recently). Ellis is, to use his own term, a *scientific essentialist*. We now need to see precisely what this position entails and what the arguments might be for or against it. (We did, of course, begin this task in Chapter 9, but now I want to look at it with a fresh eye.)

10.5 LAWS AND SCIENTIFIC ESSENTIALISM

Let's use as a working example the law that water dissolves sodium chloride. The scientific essentialist allows that, in one sense, this law might not have obtained: it might not have obtained because water and sodium chloride might never have existed. On the other hand, he maintains that the law holds of metaphysical necessity, in the sense that it obtains in every possible world in which water and sodium chloride exist (and obtains 'vacuously', perhaps, in all other worlds). Let us call the opponent of scientific essentialism the *contingency theorist*. Then the contingency theorist (or, at least, the out-and-out contingency theorist, who maintains that all laws are contingent) maintains that there are possible worlds in which water and sodium chloride exist but the law that water dissolves sodium chloride does *not* obtain. On the face of it, the contingency theorist's claim is the more plausible, because it seems that we can easily *imagine* a world in which sodium chloride simply fails to dissolve when it is immersed in water. But scientific essentialists like Ellis will emphasize at this point that imaginability is no secure guide to metaphysical possibility. For example, we may think that we can imagine a world in which Hesperus (the Evening Star) is not identical with Phosphorus (the Morning Star): and yet, if Saul Kripke is right, there is really no such possible world, given that Hesperus and Phosphorus are actually identical. According to Kripke, the identity of Hesperus and Phosphorus is *a posteriori* necessary: that is, it is metaphysically necessary, but only discoverable empirically.[8] Scientific essentialists think that, likewise, it is *a posteriori* necessary that water dissolves sodium chloride.

However, it is one thing to say this and another to argue for its truth. In the case of Hesperus and Phosphorus, Kripke himself supplies such an argument, in the form of his famous proof of the necessity of identity (a proof which is also independently creditable to Ruth Barcan Marcus).[9] What sort of argument may

---

[8] The *locus classicus* is Saul A. Kripke, *Naming and Necessity* (Oxford: Blackwell, 1980).
[9] See Saul A. Kripke, 'Identity and Necessity', in M. K. Munitz (ed.), *Identity and Individuation* (New York: New York University Press, 1971).

be mustered in the case of something like a chemical law? I can find very little in the way of any such argument in Ellis's own writings— at least, very little that is original—but in other authors we can discover rather more. For example, Sydney Shoemaker—who should be recognized as one of the first to develop an essentialist position on this question—has long maintained that properties are individuated by the causal powers which they confer upon the objects that possess them, which has the apparent consequence that laws of nature, conceived as consisting in relations between properties, must be metaphysically necessary.[10] On this view, thus, the property of being water is the very property that it is in virtue, amongst other things, of the fact that objects exemplifying it thereby possess the power to dissolve sodium chloride. In short, the reason why, on this view, there is no possible world in which water and sodium chloride exist but the former does not dissolve the latter is just that part of what it is to *be* water and sodium chloride respectively is for anything of the first kind to have a power to dissolve anything of the second kind. If that is right, then even if there is a world in which a watery-looking substance and a salty-looking substance exist, but in which the former does not dissolve the latter, this is not a world in which the first substance is water and the second is sodium chloride.

But why should we suppose that properties are 'individuated' in this way? The thought seems to be something like this. We can only tell that an object possesses a certain property by discovering whether or not the object behaves in certain ways that are characteristic of things possessing that property. For example, one way of telling that an object is round is by seeing whether it will roll down an inclined plane—although, to be more exact, a round object may only be expected to do this if it is also heavy and rigid. (Thus, a steel ball will do so, but a soap bubble will not: so, to be precise, roundness only confers upon an object which possesses it a *conditional* power to roll down an inclined plane—conditional, that is, on the object's possession of a suitable range of other properties.) There are, of course, many other ways of detecting roundness in an object, such as by measuring its cross-sections with callipers. But, it seems, we can have no grounds at all for attributing undetectable properties to physical objects and the only way in which we can detect their properties is by means of the distinctive effects which the objects possessing them are thereby able to have on other objects, including our sense organs. Consequently, it seems, the identity of a property is fixed by its causal role and thus by its place in the system of laws which actually obtains. Hence, that role and that place are essential to the property and *it* could not exist in any possible world in which the laws were different.

This sort of argument is highly dubious, I think—first because it proceeds from epistemic considerations to a metaphysical or ontological conclusion, and second

---

[10] See Sydney Shoemaker, 'Causality and Properties', in his *Identity, Cause and Mind: Philosophical Essays* (Cambridge: Cambridge University Press, 1984), and 'Causal and Metaphysical Necessity', *Pacific Philosophical Quarterly* 79 (1998), pp. 59–77.

because it conflates a question concerning the 'individuation' of properties in the actual world with a question concerning the so-called transworld identity conditions of properties. Here I shall just focus on the second issue. It is one thing to maintain that the identity of a property in this, the actual world, is fixed by its place in the actual system of laws and quite another to maintain that its having that place is an essential feature of the property, possessed by it in any possible world in which it exists. Recall from Chapter 9 the following parallel. Plausibly, no two material objects—or, at least, no two material objects of the same kind—can exist in the same place at the same time, so that the identity of a material object is fixed in this, the actual world, by its spatiotemporal trajectory (its so-called 'world line'). That is to say, if x and y are material objects of the same kind, then x and y are identical if and only if x and y have the same spatiotemporal trajectory. However, almost everyone will happily concede that any given material object could have had a different spatiotemporal trajectory from the one that it actually has, so that its actual spatiotemporal trajectory is not an essential feature of the object. Likewise, then, we may perhaps concede that if Fness and Gness are physical properties, then Fness and Gness are identical if and only if Fness and Gness enter into the same laws in the same ways, without thereby having to concede that its place in the actual system of laws is an essential feature of a physical property. At most it has been shown that different properties cannot share the same causal role in the same possible world, not that a property cannot have different causal roles in different possible worlds.

10.6 DOES WATER *NECESSARILY* DISSOLVE SODIUM CHLORIDE?

So far, I have only been looking for perfectly general arguments in support of scientific essentialism, and I shall return to this quest again shortly. But before doing so, we should briefly consider also the possibility of there being more specific arguments defending the metaphysical necessity of individual laws, such as the law that water dissolves sodium chloride. Alexander Bird, for instance, has advanced one such argument.[11] One reason why one might suppose that this law is *not* metaphysically necessary is that one might suppose that, in some possible worlds, the fundamental constants of nature have slightly different values from their actual values, with the consequence that, say, the dipole moment of water molecules is not sufficient to overcome the elecrostatic forces which maintain the lattice structure of sodium chloride crystals. However, since it is the *same* forces that are both responsible for the dipole moment of water molecules and maintain

---

[11] See Alexander Bird, 'Necessarily, Salt Dissolves in Water', *Analysis* 61 (2001), pp. 267–74, and 'On Whether Some Laws are Necessary', *Analysis* 62 (2002), pp. 257–70 (the latter being a response to criticisms published in the same issue by Helen Beebee and Stathis Psillos).

the lattice structure in question, it is very plausible that weakening or strengthening those forces somewhat would make no difference to the chemical behaviour of water and sodium chloride—that is, that in any world in which the forces are strong enough to maintain the lattice structure, they are also strong enough to confer upon water molecules a dipole moment sufficient to bring about the dissolution of sodium chloride crystals. If true, this is interesting and shows that we could indeed have scientific grounds for thinking that a certain law is metaphysically necessary. However, such reasoning would apparently only apply to the laws of special sciences, such as chemistry, and not to the fundamental laws of physics. Indeed, the reasoning in the present case presupposes that the latter laws are *not* metaphysically necessary, for it presupposes that electrons are governed by somewhat different electrostatic laws in different possible worlds, depending on slight differences in the numerical values of certain natural constants. Now, of course, philosophers and scientists do sometimes speculate about whether or not the precise numerical values of the fundamental constants of nature are themselves in any deep sense necessary. But this is, at present, no more than speculation. And although the so-called 'fine-tuning' arguments suggest that even tiny differences in the values of these constants would have resulted in physical universes radically unlike our own, this is very far from implying that those universes would have shared no properties in common with our own.

10.7 BRADLEY'S REGRESS AND THE ONTOLOGICAL STRUCTURE OF LAWS

Now I want to turn to a more solidly metaphysical kind of reason for suspecting that laws, or at least some laws, may be metaphysically necessary, one which turns on the ontological structure of laws and the problem known as Bradley's regress.[12] The problem has to do with the ontological status of the relations which 'connect' the constituents of laws and, indeed, of other facts or states of affairs. It will be recalled that I distinguish between two such relations—instantiation and characterization—whereas Armstrong is restricted to one, although I shall not dwell on this difference for the time being. Consider first, then, a singular fact or state of affairs concerning a particular object and one of its properties. Suppose, for instance, that it is a fact that a particular rubber ball is yellow. Armstrong will say that this fact consists in the ball's instantiating or exemplifying the universal yellowness. I will say, rather, that it consists in the ball's being characterized by a yellow mode or trope—that is, by a particular instance of yellowness. But what *is* this relation of 'characterization' that supposedly holds between the object and the mode? Is it a relational mode—an instance of a relational universal? We had better

---

[12] For an account of Bradley's regress, see Kenneth R. Olson, *An Essay on Facts* (Stanford, CA: CSLI, 1987), ch. 3.

not say so, for then we are heading directly for Bradley's regress. If a relational mode is needed to connect an object to one of its non-relational modes, such as a ball to its particular yellowness, then, by the same token, it would seem, two more relational modes would be needed to connect the ball to the first relational mode and this again to the particular yellowness.

The only way out of this difficulty, it seems, is to deny that the characterization 'relation' is any type of *entity* or *being* whatever, whether relational or non-relational in nature—as we saw in Chapter 3 above. What we can say instead, though, is that characterization is an *internal* relation and as such, in Armstrong's useful phrase, 'no addition of being'. In order for an internal relation to hold between two or more entities, it is sufficient for those entities to exist. A paradigm example of an internal relation, it would seem, is numerical distinctness or non-identity. Thus, the pair of objects {a, b} serves as the truthmaker of the statement that a is distinct from b. That pair of objects does not likewise serve as the truthmaker of the statement that a is five metres away from b, however, since the pair could exist and yet the statement fail to be true (unless, perhaps, a and b are points of absolute space). For this reason, distance relations (or, at least, distance relations between material objects) are external relations. Now, there is no problem in saying that the relation between an object and one of its modes—the characterization relation—is an internal relation, because it does indeed seem to be correct to say that the existence of the object and the mode suffices for the truth of the statement that the object possesses, or is characterized by, the mode. Indeed, it would seem that the existence of the *mode* suffices for the truth of that statement. This is because modes plausibly depend for their very existence and identity upon the objects which possess or 'bear' them. If a ball possesses a particular yellowness, then *that* very yellowness could not be possessed by any other object and could not exist unpossessed by any object whatever, free-floating and unattached. Hence, in every possible world in which that yellowness exists, the ball also exists and possesses that yellowness, so that in every such world and in no other it is true that the ball possesses, or is characterized by, that yellowness. Of course, none of this implies that the ball *is necessarily yellow*, because we have been given no reason to suppose that the ball could not have possessed, instead of a yellowness mode, a mode of some other colour.

So far so good, at least as far as my own ontology is concerned. But what about Armstrong's ontology, which does not include modes? What is he to say about the singular fact that a particular ball is yellow? He will say that the constituents of this fact are the ball and the universal yellowness, the latter being instantiated or exemplified by the former. But can this Armstrongian relation of instantiation be an internal relation? Seemingly not, because it apparently does not suffice, for it to be true that the ball is yellow, simply that the ball should exist and that yellowness should exist. Very plausibly, there are possible worlds in which the ball exists but exemplifies another colour universal and in which yellowness exists but is exemplified by other particulars. What Armstrong tends to say about this issue is that the lesson is that states of affairs are ontologically more fundamental than the particulars and universals that are their constituents.[13] On this view, we should not think of particulars and universals as ontologically self-subsistent entities which somehow come together in states of affairs, but rather we should think of them as being abstractions from states of affairs—as it were, 'invariants' of different types which 'run across' different states of affairs in different ways. Particulars recur in different states of affairs in one way, as being the bearers of many different properties and the relata of many different relations, while properties and relations recur in different states of affairs in another way, as being borne by many different particulars or holding between many different ordered n-tuples of particulars. However, it is not clear how this way of thinking of particulars and universals permits us to allow it to be *contingent* that a given particular possesses a given property or stands in a given relation to one or more other particulars. Indeed, in his most recent writing, influenced by the work of Donald Baxter, Armstrong seems to be gravitating towards the position that there is *no* room for contingency in this matter (even though Baxter himself is not of this opinion).[14]

Now, if there is no room for such contingency within Armstrong's ontology, how can there be room within it for contingency in the laws of nature? Recall that, for Armstrong, a law of nature is, in the simplest case, a fact consisting in two first-order universals standing in a 'second-order' relation to one another, a relation that he used to call 'necessitation'—the fact, say, that Fness necessitates Gness. But he formerly used to insist that it was *contingent* that this relation obtained between the first-order universals in question. This puzzled his critics, who claimed not to be able to understand this notion of 'contingent necessitation', which sounds almost like a contradiction in terms. Be that as it may, the deeper problem is that associated with the threat of Bradley's regress. If the law is contingent and its only constituents are Fness, Gness, and the second-order relation of necessitation, then the existence of those constituents alone is not sufficient to make the law statement true. So what must additionally exist? Armstrong once again takes the line that the law suffices for the truth of the law statement even though it has no further constituents, because the constituents should not be thought of as being ontologically prior to the states of affairs in which they figure, but as being in some sense abstractions from those states of affairs. But then, as in the case of singular facts, it becomes hard to see how laws can be contingent. And it seems that Armstrong himself is now leaning strongly towards this view and so coming round to the position of scientific essentialism. (These issues, it will be evident, are closely related to some discussed earlier in Chapter 7, in connection with Ramsey's problem.)

---

[13] See, for example, Armstrong, *A World of States of Affairs*, p. 118.
[14] See David M. Armstrong, 'How Do Particulars Stand to Universals?', in D. W. Zimmerman (ed.), *Oxford Studies in Metaphysics* (Oxford: Oxford University Press, 2004). See also David M. Armstrong, *Truth and Truthmakers* (Cambridge: Cambridge University Press, 2004).

10.8 HOW SOME LAWS MAY BE CONTINGENT

However, all is perhaps not lost for the contingency theorist, for my own ontology seems better equipped to allow for the contingency of at least some laws. By my account, a law consists, in the simplest case, in the fact that a kind is characterized by a certain property or that two or more kinds are characterized by a certain relation. Consider, thus, the law that electrons possess unit negative charge. And compare this with the singular fact discussed earlier, that a particular ball possesses, or is characterized by, a certain particular yellowness. Just as we can allow that that very ball might still have existed even if it had not possessed that yellowness, but a mode of a different colour universal, it might seem that we can allow that the kind electron could still have existed even if it had not possessed the attribute of unit negative charge, but some other attribute instead. However, the parallel is unfortunately not exact, because a contingency theorist should presumably be unwilling to say that the attribute of unit negative charge could not have existed if the kind electron had not existed, nor that it could only have existed if possessed by that kind of particle. After all, other kinds of fundamental particle do in fact possess unit negative charge in addition to electrons, and it might well seem that these particles could have existed, possessing that same charge, even if electrons had not.

The problem that we are faced with here is that attributes are not ontologically dependent upon kinds in the way that modes are upon objects. It is true that, if we are immanent realists, we shall say that *all* universals—both kinds and attributes—are, in another way, ontologically dependent upon their particular instances, inasmuch as we hold that those universals could not have existed uninstantiated. But this is a different type of ontological dependence—a species of so-called generic dependence—in that a universal does not depend for its existence upon the existence of this or that particular instance, but only upon there being *some* particular instances of it.[15] Hence, a universal also does not depend for its *identity* upon any of its particular instances, in the way that a mode depends for its identity upon the object which bears it. Furthermore, the generic dependence of universals upon their instances does not seem to help us to understand how laws could be contingent: it merely enables us to understand how the same law can obtain in worlds populated by different particulars.

Perhaps what we should say is this, however. In the case of a kind of fundamental particle, like the kind electron, perhaps the laws governing it are indeed metaphysically necessary. We can say that the reason why it is metaphysically necessary that electrons possess unit negative charge is simply that the kind electron depends for its existence and identity upon the attribute of unit negative

---

[15] See further my *The Possibility of Metaphysics*, p. 141, and the discussion of 'non-rigid' existential dependence in Chapter 3 above.

charge—although equally upon certain other attributes, such as a certain characteristic rest mass and a spin of one half. It will still be the case that *modes* of those attributes, belonging to particular electrons, will depend for their existence and identity upon those electrons. Thus, on this view, the dependency relations obtaining at the level of universals will be exactly the reverse of those obtaining at the level of particulars, at least in the case of kinds of fundamental particle (recall, here, Fig. 7.2 in section 7.8 of Chapter 7 above). Then, just as the existence of a particular electronic charge—a mode—is a truthmaker of the statement that a certain electron bears that charge, so the existence of the kind electron will be a truthmaker of the law statement that electrons possess unit negative charge. Even so, this does not have the consequence that the same *attribute* cannot exist in other possible worlds, being possessed in those worlds by different kinds of object. We can still allow, thus, that unit negative charge exists in other possible worlds and is possessed in those worlds by kinds of fundamental particle which possess, along with unit negative charge, other attributes which are either not possessed at all by anything in the actual world or are not possessed in the actual world by anything in conjunction with unit negative charge. These, then, would be kinds of fundamental particle which do not exist at all in the actual world. This being so, we are already in a position to cite lawlike truths which are, plausibly, metaphysically contingent. For the purpose of the example, I take it that positrons differ from electrons only in being positively rather than negatively charged and that the kind electron has just the following three attributes: unit negative charge, rest mass m, and spin one half. Then it is a lawlike truth in the actual world, but quite feasibly not in some other possible worlds, that any kind of particle with rest mass m and spin one half has unit (positive or negative) charge.

But what about the intuition, which many contingency theorists will have, that electrons—that very kind of particle—could have had at least a slightly different quantity of charge associated with them? (It may be readily conceded that electrons couldn't have been *positively* charged, because this seems to be tantamount to saying, absurdly, that electrons could have been *positrons*.) Well, perhaps saying that it is metaphysically necessary that electrons have unit negative charge is not, in fact, tantamount to saying that they couldn't have had a slightly different quantity of negative charge, because saying that they could have had a slightly different quantity of negative charge might be interpreted as implying that the *unit* of negative charge could have been slightly different from what it actually is. Indeed, this seems to me very plausible. So this is another way in which we might be able to allow for contingency in the laws of nature, even at the level of fundamental physics. And in the case of kinds of object at higher levels of complexity, there seems to be even more room for contingency in laws—for instance, in the case of laws concerning chemical kinds. It is widely supposed that water is essentially hydrogen oxide—H_2O—and I shall not question that supposition here. Equally, it may be said that hydrogen is essentially that element whose atomic nucleus contains a single proton, which may or may not be accompanied by one or even two neutrons. And protons, we may concede, possess unit positive charge of metaphysical necessity. But all of these facts may be consistent with supposing that water—that very stuff—might have behaved chemically in a somewhat different fashion, or is governed by somewhat different chemical laws in other possible worlds. It may be that Bird's line of argument works in the case of the solubility of sodium chloride in water, but this is only one case and not obviously generalizable to cover all other chemical laws involving water. The more complex a kind of object is, the less evident it is that metaphysical necessity governing the laws of fundamental physics—assuming it to obtain—must be transmitted upwards to laws at this higher level of complexity. The upshot is that, although it seems difficult to avoid the conclusion that *some* laws must be metaphysically necessary, it is far from evident that all laws must be so.

10.9 STRUCTURE-BASED LAWS

Here it may be objected that if the argument based on the threat of Bradley's regress is any good at all, and can be used to sustain the conclusion that certain fundamental laws are metaphysically necessary, how can it fail to sustain the same conclusion for other, less fundamental laws, such as chemical laws? The reason is, at bottom, that less fundamental laws govern complex kinds of objects, whose complexity consists in the fact that such objects are composed of many other objects standing in complex structural relationships to one another. These structural relationships will involve external relations, such as distance relations and force relations. Thus, the truthmakers for statements ascribing attributes to such kinds of objects will involve facts about such external relations, not just facts whose only constituents are fundamental kinds and their attributes. (For example, the law that water dissolves sodium chloride is, on closer investigation, a law concerning a type of chemical interaction between hydrogen oxide molecules and sodium chloride crystals, and both of these chemical kinds are so-called 'structural' universals: the kind *hydrogen oxide molecule*, thus, comprises the kinds *hydrogen ion* and *oxygen ion* standing in that complex structural relation which gives all of its particular instances the structure of two hydrogen ions each bonded covalently with the same oxygen ion.) Consequently, we cannot simply generalize the style of argument based on the threat of Bradley's regress so as to apply it directly in the case of higher-level laws governing the behaviour of complex kinds of objects. It is certainly open to the opponents of contingency theorists to attempt to extend this style of argument if they can, but the onus of proof is undoubtedly upon them.

It is worth remarking, in this context, that physical structures are notoriously sensitive to variations of scale: if a human body were quadrupled in size, it would not be biologically viable. There is every reason to suppose, thus, that certain biological laws depend upon the size of the kinds of organisms to which they apply. But the size of a kind of organism is itself to some degree contingent upon its evolutionary history. Primitive horses were much smaller than modern ones but it is conceivable that, had evolutionary pressures been different, horses might have evolved to become even smaller instead of larger. This would assuredly have had a significant impact upon the laws of equine behaviour. I cite this example only for purposes of illustration of a general principle, which once again lends credence to the claim of the moderate contingency theorist that at least some laws are not metaphysically necessary.

10.10 THE QUESTIONABLE NOTION OF NOMIC NECESSITY

So let us return to the issue with which we began. *Is* there a species of necessity—natural or nomic necessity—which flows from the laws of nature and is somehow weaker than logical or metaphysical necessity? Certainly, we can define some event or state of affairs as being 'naturally necessary' just in case it occurs or obtains in every possible world in which the actual laws of nature hold. But, first, it is not clear what class of events or states of affairs this definition will capture (although it will, trivially, include all of the actual laws of nature themselves) and, second, not all of these events or states of affairs will have the same modal status as far as the logical or metaphysical modalities are concerned. With regard to the second point: some of these events or states of affairs will be metaphysically contingent, some will be metaphysically necessary for reasons which have nothing to do with the laws of nature, and some will be metaphysically necessary on account of the metaphysical necessity of certain natural laws. With regard to the first point: our analysis of laws implies that many laws have implications only for the causal powers and liabilities of particular objects, not for their actual behaviour. The law that water dissolves sodium chloride implies that any particular body of water has a power to dissolve sodium chloride and any instance of the law is a particular case of some water's dissolving some sodium chloride. But the law can hold in a world containing water and sodium chloride without there necessarily being any particular case of some water's dissolving some sodium chloride, that is, the law can hold in a world even though it has no instances in that world. (This is no counterexample to the thesis of immanent realism, because laws are not themselves universals, even though they involve universals, which must be instantiated according to immanent realism: the sense, thus, in which a law has 'instances' is different from that in which a universal has 'instances', and ideally we should use different terms to express these senses. We could speak, thus, of 'cases' of a law.)

There are many reasons why the law that water dissolves sodium chloride might hold in a world containing water and sodium chloride without there being in that world any particular case of some water's dissolving some sodium chloride. It might be, for instance, that none of the water in that world ever comes into contact with any of the sodium chloride, either as a matter of pure chance or somehow as a consequence of the world's arrangement at the beginning of time—its 'initial conditions'. Alternatively, it might be—again, possibly by accident—that whenever some water comes into contact with some sodium chloride in this world, some other chemical or physical agent is also present which interferes to prevent the chemical interaction between them that would otherwise have taken place. Determinists may believe that every event that occurs in a world occurs as a necessary consequence of that world's initial conditions and the laws of nature that hold in it. But it is far from clear that this is the correct way to see the relationship between the actual behaviour of physical objects and the laws of nature. Laws, it seems, determine the tendencies or propensities of physical objects to behave in this or that way, but do not determine—even in conjunction with 'initial conditions'—every detail of their actual behaviour. At least, the assumption that the latter is the case should be seen as a highly controversial one and not simply accepted unquestioningly: a fact that is now increasingly being acknowledged by philosophers of science, thanks to the work of Nancy Cartwright.[16]

On balance, I think it is best to abandon the idea of there being any distinctive species of 'natural' or 'nomic' necessity, on the grounds that there is no reason to think that such a notion could capture a homogeneous class of events or states of affairs all of which have a modal status in common which is somehow 'weaker' than logical or metaphysical necessity. This, as we can see, is not to abandon the notion of natural law itself, nor is it to insist with the scientific essentialists that natural necessity just reduces to metaphysical necessity. I return to my earlier claim, that many and perhaps even all natural laws just *are* facts about the natures of certain natural kinds of things or stuff. And by the 'nature' of a natural kind of things or stuff, I simply mean its attributes—its properties and relations, conceived as universals. In the case of very fundamental kinds of things, such as electrons, their nature may be fairly simple, comprising just a few intrinsic properties. But the world contains also many very complex things, with correspondingly complex natures which may never be completely known to us. Although all of a kind's nature may be essential to it in the simplest cases, such as that of the kind electron, it is far from clear that this must be so in the case of complex kinds, such as chemical and biological kinds. Consequently, as we have seen, not all of the laws of nature need be metaphysically necessary, although some of them almost certainly are.

---

[16] See Nancy Cartwright, *Nature's Capacities and Their Measurement* (Oxford: Clarendon Press, 1989), and *The Dappled World: A Study of the Boundaries of Science* (Cambridge: Cambridge University Press, 1999).

PART IV

TRUTH, TRUTHMAKING, AND METAPHYSICAL REALISM

11

Metaphysical Realism and the Unity of Truth

11.1 METAPHYSICAL REALISM AND ALETHIC MONISM

Many forms of dualism or pluralism are apparently compatible with metaphysical realism—that is, with the doctrine that things and states of affairs generally exist independently of their being objects of thought. For example, dualism in the philosophy of mind is compatible with such realism, as is theological pluralism—the doctrine that there are many divine beings. But one fundamental form of monism to which metaphysical realism is apparently committed is what might be called *alethic* monism: a conception of truth which holds truth to be unitary and indivisible. Such a conception of truth is opposed to all relativist views of the nature of truth, whether these spring from philosophical idealism or from one or other form of cultural relativism. My aim in this chapter will be to defend metaphysical realism by advocating alethic monism. An important part of that defence will consist in demonstrating the tight connection between the two doctrines.

11.2 TRUTH-BEARERS AND THEIR MULTIPLICITY

How could there be many truths, rather than just one truth? In one sense, of course, it is relatively uncontentious that there are many truths, if one thinks that there are any truths at all. This is the sense in which there are many different *truth-bearers*, each of which bears the property (if property it is) of being true. What the truth-bearers *are* is a further and difficult question. On some views, they are linguistic entities, such as sentence-tokens or sentence-types of one or more actual or possible languages. On other views, they are propositions, conceived as non-linguistic entities which may, however, constitute the meaningful contents of sentential expressions or the intentional contents of such mental states as beliefs, desires, hopes and fears (the so-called propositional attitudes). And there are, of course, various different theories as to the nature of propositions themselves. Some theorists hold propositions to be wholly abstract entities, having an ontological status akin to that of Platonistically conceived mathematical objects. Others hold propositions to be structured complexes which may contain, amongst their constituents, concrete physical objects and their properties or relations. According to the former approach, the proposition that Mont Blanc is snow-capped may represent something about the condition of that physical object, but it does not literally contain the mountain itself as a constituent, as the latter approach would hold.[1]

If we are to talk about truth, we must make some decisions about the nature of truth-bearers. But we needn't be monistic about truth-bearers: we needn't maintain that all truth-bearers belong to a single, unified ontological category. We can be pluralistic and allow that, in their own ways, such diverse entities as sentences, beliefs, and propositions may qualify as truth-bearers. But isn't there then a danger of truth itself being fragmented—a danger that we shall have to operate with different conceptions of truth when speaking of the various different kinds of truth-bearer? Not necessarily, so long as we can maintain that one kind of truth-bearer has primacy: that there are 'primary' as well as 'secondary' truth-bearers. For then we can regard truth in the primary sense as a property of the primary truth-bearers and define truth in various secondary senses in terms of primary truth. For instance, we can say that truth in the primary sense is truth of a *proposition* and define the truth of a sentence as the truth of the proposition that it expresses and the truth of a belief as the truth of the proposition that constitutes its intentional content.

And, in fact, it would seem that this is our most promising option. If we held the truth of sentences, say, to be primary, then we would face difficulties in defining the truth of either propositions or beliefs, because, very arguably, there can be true propositions and true beliefs that cannot be expressed in any actual language. Suppose, for instance, that we said that a proposition or a belief is true if and only if it is expressed by some true sentence. Then, in the first place, we face the difficulty that there is, plausibly, an uncountable infinity of true propositions, but not even a countable infinity of actually existing sentences. In the second place, we face the difficulty that there may be creatures that have true beliefs which they are incapable of expressing in language and which are not translatable into any actually existing language. Perhaps we could get around these difficulties by appealing not to actually existing sentence-*tokens* (particular sentential utterances or inscriptions) but to abstract sentence-types of merely possible languages, including infinitary languages with a countable infinity of distinct expression-types and syntactical rules allowing the infinite concatenation of such expression-types. However, then it is difficult to see how our ontology of primary truth-bearers is going to differ, in any really significant way, from that which takes propositions, conceived as abstract entities, as being the primary truth-bearers.

---

[1] These two different approaches are exemplified by the views of Gottlob Frege and Bertrand Russell respectively, as is brought out in a famous exchange of letters between them involving this very example of Mont Blanc and its snowfields: see Gottlob Frege, *Philosophical and Mathematical Correspondence*, ed. Brian McGuinness, trans. Hans Kaal (Oxford: Blackwell, 1980), p. 163 and p. 169, reprinted as 'Selection from the Frege–Russell Correspondence', in Nathan Salmon and Scott Soames (eds), *Propositions and Attitudes* (Oxford: Oxford University Press, 1988), pp. 56–7.

Another possibility, for those who seek to make linguistic entities the primary truth-bearers, might be to appeal to what has been called a 'Lagadonian' language, named after Jonathan Swift's amusing account in *Gulliver's Travels* of a (fictitious) people who attempted to use the very things referred to as 'words' for themselves—so that, for example, Mont Blanc (that very mountain) would, in their language, be the name of itself.[2] Clearly, such a 'language' would not really be a *language* in the familiar sense of the term, although that in itself constitutes no objection to the proposal to treat its 'sentences' as being the primary truth-bearers. More to the point, however, it is still not clear that this strategy would serve to overcome the problem about the expressive limitations of language: for there are, very arguably, infinitely many truths about what is *possibly* the case that could not be expressed in a 'Lagadonian' language, simply because they are truths which do not purely concern entities which *actually* exist. Moreover, even a 'Lagadonian' language would, it seems clear, need to include abstract objects in the form of set-theoretical constructions, so that appealing to such a language would provide no way of restricting the primary truth-bearers to purely concrete entities.[3]

Anyway, without more ado, I am going to suppose that propositions, conceived as abstract objects, are the only suitable entities to play the role of primary truth-bearers, on the grounds that propositions (or entities very like them) are alone sufficiently numerous to constitute all the truths there are. Here, however, it may be inquired why we have to assume, in any case, that there are indeed uncountably many truths. Why can't we just say, for example, that there are as many truths as there are true sentences that have at some time or other been uttered? The trouble is that truth is closed under entailment, so that anything entailed by one or more truths is itself true. Thus, we know that the set of all true sentences that have at some time or other been uttered cannot collectively exhaust the truth, because these truths entail further truths—infinitely many of them—that have never been uttered and never will be. To this it may be objected that we cannot rule out the possibility that the physical universe has an infinite future, in which, at some time or other, every one of these infinitely many truths will be uttered. However, we know *a priori* that these infinitely many truths exist, whereas we do not know *a priori* that infinitely many true utterances will be made in the future, from which it seems we may conclude that those utterances cannot be the primary truth-bearers: it will at best be a cosmic accident if it should so happen that, in the course of an infinite future, every truth is at some time uttered. Moreover, this ignores the problem that there are truths which could only be expressed in a language by means of infinitely complex sentences—and such sentences will never be uttered, even in the course of an infinite future, because it is impossible for them to be uttered.

---

[2] For the notion of a 'Lagadonian' language and complaints about its expressive limitations, see David Lewis, *On the Plurality of Worlds* (Oxford: Blackwell, 1986), pp. 145 ff. For doubts about those complaints, see Joseph Melia, 'Reducing Possibilities to Language', *Analysis* 61 (2001), pp. 19–29.

[3] As Melia describes the ontological commitments of such a language, it 'postulates only elements of the concrete world and set-theoretic constructions thereof': see 'Reducing Possibilities to Language', p. 20.

Against my contention that abstract propositions are the primary truth-bearers it may be objected that such objects are *entia non grata* precisely because they are abstracta. If truth requires the existence of abstract entities, so much the worse for truth, it may be said. This might be a fair objection if it could be satisfactorily explained what, precisely, is wrong with admitting abstract entities into one's ontology. The usual objections to abstracta are, on the one hand, epistemological in character and, on the other, metaphysical. On the epistemological side, it is commonly objected that, assuming a causal theory of knowledge and given that abstract objects are causally inert, we could have no knowledge of abstracta. (A causal theory of knowledge is one which holds, roughly, that knowledge about an object consists in belief which is reliably caused by that object: for instance, that I know that the sun has risen because my belief that it has risen has been caused via a reliable perceptual mechanism by the rising of the sun.) Accepting that abstract objects are causally inert, however, we may simply reject causal theories of knowledge as adequate for anything but empirical knowledge of concrete objects.[4] Knowledge of propositions is *a priori* knowledge, like knowledge of mathematical objects. This is not, of course, to say that knowledge of the *truth or falsehood* of propositions is always *a priori* knowledge—just knowledge of their existence and of their intrinsic properties and relations to one another, including their entailment relations.

As for the metaphysical objections to abstracta, these are, if anything, even feebler than the epistemological objections. The objections typically arise from dogmatic forms of extreme naturalism, which hold that only physical things existing in space and time exist at all. This doctrine has no scientific foundation. Certainly, it has no foundation in physical science, since physical science is only qualified to inform us concerning what physical things exist, not concerning what non-physical things do or do not exist. If we are then challenged to say why we need to suppose that non-physical abstracta exist at all, our answer can simply be that we know many things to be true which can only be true if abstracta exist, such as many of the truths of logic and mathematics. And, for reasons just given, we need not feel at all embarrassed to claim that we *know* these things to be true, *a priori*: for the possibility of such *a priori* knowledge is not seriously threatened by causal theories of knowledge.

11.3 THE INELIMINABILITY OF TRUTH

Rather than concurring with the sentiment that 'if truth requires the existence of abstract entities, so much the worse for truth', I would urge, rather, the contrary sentiment that 'since truth requires the existence of abstract entities, so much the better for abstract entities'. This, of course, presupposes that we do indeed need to acknowledge an ineliminable role for truth in our intellectual life. Do we, though?

---

[4] Compare Jerrold J. Katz, 'Mathematics and Metaphilosophy', *Journal of Philosophy* 99 (2002), pp. 362–90.

Notoriously, we do not if some sort of 'redundancy' theory of truth is acceptable. So that is something that we need to look at before proceeding further. Since I am assuming that abstract propositions are the primary truth-bearers, it would be appropriate, at least initially, to consider a version of the redundancy theory which likewise has this starting point.

An adherent of such a version of the theory will maintain that to assert, for example, that the proposition that Mars is red *is true* is equivalent to asserting the proposition that Mars is red. This may be granted. It doesn't follow, of course, that the concept of truth, as it figures in the assertion that the proposition that Mars is red is true, is simply vacuous and so makes no substantive contribution to what is asserted. The case is not even akin, say, to that of the equivalence between asserting that Mars is red and asserting that Mars is red *and* Mars is red. In fact, what the former equivalence reveals, if anything, is not the redundancy of the concept of truth so much as something constitutive of assertion: that assertion aims at truth. To *assert* that Mars is red is precisely to represent the proposition that Mars is red as being *true*. And since, if a proposition is true, then it is true that it is true, it should come as no surprise that to assert a proposition is equivalent to asserting its truth.

In any case, the redundancy theorist must, of course, at least be able to show how expressions of truth are 'eliminable' in all possible linguistic contexts. And it is highly questionable whether this can be shown. Take the familiar example of an assertion that Pythagoras's Theorem is true. The redundancy theorist may urge that 'Pythagoras's Theorem is true' is equivalent to something like this: 'For any proposition p, if Pythagoras's Theorem is that p, then it is true that p'. And he may then urge that this in turn is equivalent to 'For any proposition p, if Pythagoras's Theorem is that p, then p', on the grounds that, just as 'It is true that Mars is red' is equivalent to 'Mars is red', so, quite generally, 'it is true that p' is equivalent to 'p', whatever proposition p may be. But there are many familiar problems with this line of argument. One is that it is, strictly, nonsense to say that, for any proposition p, 'it is true that p' is equivalent to 'p', because the quotation marks here play havoc with what the theorist is trying to maintain. To put the point in technical terms, one cannot intelligibly *quantify into* a context sealed off by quotation marks. Hence, the general principle to which the redundancy theorist is trying to appeal must be stated, somehow, without the use of quotation marks. He might, for instance, attempt to state it thus: for any proposition p, it is true that p is equivalent to p. But this is equally unintelligible, because it is not even grammatically well-formed (unless the clause following the quantifier is parsed, inappropriately, as 'it is true that: p is equivalent to p'). Although I shall not attempt to prove it here, there simply doesn't appear to be a satisfactory way for the redundancy theorist to express the general principle he needs in order to 'eliminate' all expressions of truth without loss of informational content or expressive power.[5]

---

[5] For further discussion, see Susan Haack, *Philosophy of Logics* (Cambridge: Cambridge University Press, 1978), pp. 127 ff.

11.4 TRUTHMAKERS AND THEIR MULTIPLICITY

The position that we have now arrived at is this. Reference to truth is ineliminable from our discourse without crippling its expressive power. And the primary bearers of truth have to be abstract propositions, or something sufficiently akin to them to make no difference for our purposes. However, we still seem to be a long way from any thesis of the necessary unity of truth. Indeed, so far I have been emphasizing the necessary plurality—in fact, the uncountable infinity—of *truths*. To move closer to our goal, we need to concentrate now not, as we have been so far, on the truth-*bearers* but on the truth*makers*.[6] However, the truthmakers are also, plausibly, many rather than one. This is not, I concede, the view of all objectivist theorists of truth. According to some truth-theories, all true propositions are true in virtue of referring to the same thing, *the True*, while all false ones refer to *the False*.[7] This is monism about truth indeed, but arguably too monolithic a monism to be metaphysically acceptable. What makes the proposition that Mars is red true is, surely, *something different* from what makes the proposition that Venus is white true.[8]

Here one may be strongly inclined to say that these two different things are, respectively, *the fact* that Mars is red and *the fact* that Venus is white. However, then we are in danger of deluding ourselves about the ontological basis of truth. We are in danger, that is to say, of simply inventing appropriate truthmakers in the image of the truth-bearers to which they are supposed to 'correspond'. As others have pointed out, 'facts' seem to be too similar to propositions—mere 'shadows' of them—to constitute their truthmakers in any illuminating sense. Certainly this appears to be so on a conception of truthmaking which sees truth as consisting in a one-to-one correspondence between propositions as truth-bearers and facts as their truthmakers. Indeed, on this conception of facts, it is not clear that a fact can be understood as anything other than a true proposition, in which case it is trivial to say that a proposition is true if and only if it has a fact corresponding to it.

---

[6] Introduction of the term 'truthmaker' has been attributed to C. B. Martin: see D. M. Armstrong, 'C. B. Martin, Counterfactuals, Causality, and Conditionals', in John Heil (ed.), *Cause, Mind, and Reality: Essays Honoring C. B. Martin* (Dordrecht: Kluwer, 1989), pp. 7–15, at p. 9. Although I shall be assuming, with Martin, that truths need truthmakers, this is not question-begging in the context of a defence of metaphysical realism, since the assumption by itself is not incompatible with metaphysical idealism, subjectivism or relativism.

[7] This is Frege's view: see his 'On Sense and Reference', in Peter Geach and Max Black, *Translations from the Philosophical Writings of Gottlob Frege*, 2nd edn (Oxford: Blackwell, 1960).

[8] It is sometimes objected against the view that predicative truths like these need truthmakers that it indulges in unwarranted reification. As David Lewis has put it, 'predications . . . seem, for the most part, to be true not because of *whether* things are, but because of *how* things are': see his critical notice of D. M. Armstrong's *A Combinatorial Theory of Possibility*, *Australasian Journal of Philosophy* 70 (1990), pp. 211–24, at p. 216. See also David Lewis, 'Truthmaking and Difference-Making', *Nous* 35 (2001), pp. 602–15, and Julian Dodd, 'Is Truth Supervenient on Being?', *Proceedings of the Aristotelian Society* 102 (2002), pp. 69–86. However, as we shall see later, *how* things are is plausibly a matter of what *ways* they are, where 'ways' are construed as particular properties or 'tropes' and as such genuine entities that need to be included in our ontology.

Against the foregoing conception of facts, it may also be urged that facts had better not be mere abstracta, like propositions themselves, if they are to constitute the truthmakers of true propositions: they had better be, rather, pieces of the concrete world, at least in the case of facts which answer to propositions concerning concrete reality. This then leads us to an alternative conception of facts, according to which facts are 'complexes' which, at least in many cases, contain concrete objects and their properties or relations as constituents—a view which, of course, closely resembles the alternative conception of *propositions* briefly mentioned earlier. What we have, then, are two rival conceptions of both propositions and facts—one which represents these items as being pure abstracta and another which represents them as being 'complexes' which help to make up concrete reality. But if we adopt the *same* conception of both propositions and facts, we are left with inadequate resources wherewith to construct a substantive account of truth-making. If truthmaking consists in a relation between propositions and facts, the items thus related must be conceived of as belonging to quite different ontological categories. And since facts must obviously occupy the 'worldly' side of the relation, they had better be understood as 'complexes', while the propositions are regarded as mere abstracta. This, then, gives us an additional reason to adopt the abstract conception of propositions.

None of this yet gives us a compelling reason to include 'facts', conceived as 'complexes', in our ontology, just a reason to regard them as complexes rather than as abstracta if they are to be assigned a truthmaking role. I have already pointed out that to assign to facts such a role is to abandon a monolithic form of monism about truthmaking: it is to abandon the view that it is *the same thing*—the True or, perhaps, the world as a whole—that makes all true truth-bearers true. However, to assign the truthmaking role exclusively to facts is still to adopt a weaker form of monism about truthmaking, for it is to adopt the view that all truthmakers belong to the same ontological category. Why should we suppose this to be the case, though? Perhaps the supposition provides another example of the way in which the structure of language can prejudice our thoughts about ontology. Our canonical way of referring to propositions is by means of the sentences that we take to express them: we refer to a certain proposition as, for example, the proposition that Mont Blanc is snow-capped. And in an analogous fashion, our canonical way of referring to properties is by means of the predicates that we take to express them: we refer to a certain property as, for example, the property of being snow-capped. It is very natural to suppose, then, that what makes the proposition that Mont Blanc is snow-capped true is some 'complex'—a fact—which contains as its 'constituents' both a certain object, Mont Blanc, and a certain property, the property of being snow-capped. This supposed entity is something which, once again, it seems we can most perspicuously refer to by means of the very same sentence through which we picked out the proposition at issue: and thus we refer to it as 'the fact that Mont Blanc is snow-capped'. And so it is for other propositions—except that we need not necessarily suppose that each true proposition requires a distinct fact as its truthmaker, giving rise to a one-to-one correspondence between true propositions and facts: for we can maintain, for instance, that there are no *disjunctive* facts needed to serve as the truthmakers of disjunctive propositions, nor any *negative* facts needed to serve as the truthmakers of negative propositions. Instead we can say that a true disjunctive proposition is made true by any fact which makes either of its disjuncts true and that a negative proposition is derivatively true in virtue of the non-existence of any fact which would make the corresponding positive proposition true. While this ontological economy is to be welcomed, however, we still haven't been given a clear reason to suppose that all true propositions are made true by entities all of which belong to a single ontological category—the category of 'facts'.

11.5 FACTS AND TRUTHMAKING

One line of argument which might seem to remedy this deficiency is the following.[9] Take once more the true proposition that Mont Blanc is snow-capped. And make at least this assumption about the nature of truthmaking: that a truthmaker makes a proposition true just insofar as the *existence* of the truthmaker entails the truth of the proposition. (This assumption is not entirely uncontroversial, as I shall acknowledge more fully shortly, but let us go along with it for the time being.) Then it emerges, in the case of our example, that the truthmaker of the proposition that Mont Blanc is snow-capped must be something whose existence entails the truth of that proposition. Clearly, it would seem, that thing cannot simply be Mont Blanc itself, nor can it simply be the property of being snow-capped, nor can it even be, more esoterically, the ordered pair whose first member is Mont Blanc and whose second member is the property of being snow-capped, <Mont Blanc, being snow-capped>. For it is possible for each of these things to exist in a world in which it is *not* true that Mont Blanc is snow-capped. And yet, if the thing which is the truthmaker of the proposition that Mont Blanc is snow-capped is to be something whose existence entails the truth of that proposition, then, clearly, it must be something which is such that that proposition is true in every possible world in which that thing exists. This then leads us naturally to the thought that the truthmaker of this proposition must be a 'complex' which contains both Mont Blanc and the property of being snow-capped, but not just in the 'disconnected' way in which the ordered pair <Mont Blanc, being snow-capped> contains them as members. Rather, the complex must be something which somehow consists in the property of being snow-capped *belonging to*, or being *exemplified by*, the object Mont Blanc. At the same time, it clearly will not do to represent this complex as simply being an ordered *triple*, whose members are Mont Blanc, the

---

[9] Compare D. M. Armstrong, *A World of States of Affairs* (Cambridge: Cambridge University Press, 1997), pp. 115–16.

property of being snow-capped, and the relation of belonging or exemplification, for this triple too can exist in a possible world in which it is not true that Mont Blanc is snow-capped. We see, thus—or so it seems—that the sort of 'complex' that we have to be talking about here is something which involves an entirely distinctive form of 'composition': and thus we arrive at the conception of a 'fact' as just such a complex and as being the only sort of entity that is ontologically suited to occupy the truthmaking role. This, then, supposedly motivates the kind of 'non-monolithic monism' about truthmaking that we have been examining.

I mentioned earlier, parenthetically, that the foregoing conception of the truthmaking relation is not entirely uncontroversial. According to that conception, what it is for a truthmaker to 'make true' a proposition is for the *existence* of the truthmaker to entail the truth of the proposition. Against this, it is sometimes objected that entailment, strictly speaking, is a relation between *propositions*, so that if we are to persist in this view of truthmaking, then we need more properly to say that what it is for a truthmaker to 'make true' a proposition is for the *proposition that the truthmaker exists* to entail the truth of the proposition in question. But now we seem to be faced with further and insuperable difficulties. First of all, it is surely not enough, for a proposition to be made true, that its truth should be entailed by another proposition: that other proposition also needs to be *true*. And then we seem to have introduced a fatal circularity into our account of truthmaking, because we are now purporting to explain the truth of some propositions by appealing to the truth of others. Moreover, our account of truthmaking seems to have turned truthmaking into a relation *between propositions*, that is, between truth-*bearers*, when what we were seeking was a relation between truth*makers* and truth-bearers, conceived as entities belonging to two quite different ontological categories.

I think that the solution to this difficulty—or part of the solution, at any rate—is to set aside as a verbal quibble the objection to the use of the verb 'entail' in the proposed account of truthmaking. Let it be granted that, strictly speaking, entailment is a relation between propositions: and then let us employ, instead of the notion of entailment, the notion of *metaphysical necessitation*. That is to say, let us say that what it is for a truthmaker to 'make true' a proposition is for the existence of the truthmaker to *metaphysically necessitate* the truth of the proposition, explaining this in the following way: the existence of the truthmaker metaphysically necessitates the truth of the proposition just in case *in every possible world in which the truthmaker exists, the proposition in question is true*. It cannot now be objected that this account turns truthmaking into a relation between propositions, nor that it is viciously circular. There may, of course, be further objections that can be raised against such an account of truthmaking—for instance, concerning the truthmakers of necessarily true propositions—but for present purposes I shall set those aside as relatively minor compared with what appeared to be the devastating criticism that we have just managed to defuse. I shall assume, then, that the modified account of truthmaking just proposed is basically sound, even if it may stand in need of further refinements.[10] Thus reassured, let us return to the issue of the nature of the truthmakers.

One immediate objection to the doctrine that all truthmakers are facts, even if it is restricted to contingent truths, is that contingent *existential* propositions do not need facts as truthmakers. Consider, for instance, the true contingent existential proposition that Mont Blanc exists. What is it whose existence metaphysically necessitates the truth of this proposition? The answer is simple and obvious: *Mont Blanc*. And this is an *object*, not a fact. Of course, if one believes in facts at all, one may well believe that there is such a fact as the fact that Mont Blanc exists. And then one may hold that it is this fact that is the truthmaker of the proposition that Mont Blanc exists, on the grounds that the existence of this fact metaphysically necessitates the truth of that proposition. And, of course, this fact will exist in all and only those worlds in which Mont Blanc exists, explaining why it is that the existence of Mont Blanc likewise metaphysically necessitates the truth of the proposition that Mont Blanc exists. However, this at best shows that the fact that Mont Blanc exists is a truthmaker of the proposition in question *in addition to* Mont Blanc itself, not that the latter is *not* a truthmaker of that proposition. So we are still left with the conclusion that facts are not the only truthmakers.

A more serious objection to the thesis that facts are the only truthmakers is, however, the following, which challenges the very line of argument that I presented earlier as motivating the appeal to facts as truthmakers. That argument, it will be recalled, pointed out that neither Mont Blanc, nor the property of being snow-capped, nor even the ordered pair of these two entities can qualify as the truthmaker of the proposition that Mont Blanc is snow-capped, because there are many possible worlds in which those objects exist but the proposition in question is not true. However, this presupposes that we are conceiving the property of being snow-capped here as being a *universal*, which may be exemplified by many different objects and which at most has to be exemplified by *some* object in any possible world in order to exist in that world. (If we are Platonists about universals, rather than Aristotelians, we shall not even insist that the universal needs to be exemplified in order to exist in any possible world and, indeed, will probably want to say that it exists in every possible world.) But properties need not necessarily be conceived as being universals: another way to think of properties is to think of them as being *dependent particulars* of a special kind, often referred to nowadays as 'tropes' and in former times as 'individual accidents' or 'modes'. According to this manner of conceiving properties, a property is a particular 'way' something is, as we saw in Chapter 6 above—for instance, a red object's property of being red is its particular way of being coloured. Properties so conceived are particulars, so that two red objects which are exactly similar in respect of their colour possess, none

---

[10] For some proposed refinements, see Greg Restall, 'Truthmakers, Entailment and Necessity', *Australasian Journal of Philosophy* 74 (1996), pp. 331–40, and Barry Smith, 'Truthmaker Realism', *Australasian Journal of Philosophy* 77 (1999), pp. 274–91. I develop my own account of truthmaking in ch. 12 below.

the less, two numerically distinct 'rednesses', rather than—or, indeed, *as well as*—exemplifying one and the same universal redness. Furthermore, properties so conceived are *ontologically dependent* entities, in the sense that they cannot, of metaphysical necessity, exist other than as properties of the particular objects which possess them. Thus, the particular redness of a given red object could not have been possessed by any other object, nor could it have existed without being possessed by any object at all.

Returning now to the example of Mont Blanc, we see, then, that if Mont Blanc's property of being snow-capped is regarded as being one of that mountain's tropes, or individual accidents, then it is far from clear, after all, that a 'fact' is needed to constitute the truthmaker of the proposition that Mont Blanc is snow-capped: it seems, indeed, that Mont Blanc's property of being snow-capped will serve perfectly well as the truthmaker of that proposition, because in any possible world in which that particular property exists, it is true that Mont Blanc is snow-capped. This is because, as we have just seen, that property, conceived as being a trope, cannot exist in any possible world in which Mont Blanc does not exist and cannot exist as the property of any other object in any possible world. Hence, in any possible world in which that property exists, Mont Blanc itself exists and has the property, so that in any such world the proposition that Mont Blanc is snow-capped is true.

Whether or not we favour an ontology which includes tropes as well as, or instead of, universals, the foregoing considerations should be instructive, for they demonstrate that it is far from indisputable that facts are needed as truthmakers in any metaphysically realist theory of truth. They demonstrate at the same time that a weakly monistic account of truthmakers, which holds all truthmakers to belong to the same ontological category, is also far from indisputable. A very important feature of the considerations which lead to these conclusions is that they turn on the existence of *necessary connections* between entities of different ontological categories—for instance, between objects and their particular properties. Some philosophers, of course, follow the lead of David Hume in denying the existence of any such necessary connections, holding that, as Hume would put it, all existing things are distinct and separable, or capable of existing independently of one another. However, this is a difficult doctrine to defend and will not serve the friends of facts particularly well.[11] For, as we have seen, facts have to be conceived of as being complexes which are composed by their 'constituents' in a very special sort of way, if we want to attribute to facts the strong kind of unity they need in order to function as truthmakers. If we take the fact that Mont Blanc is snow-capped to contain the constituents Mont Blanc and the universal property of being snow-capped, standing to one another in the 'exemplification' relation, then, plausibly, this is not the sort of entity that will comfortably find a place in a Humean ontology

---

[11] David Lewis, who favours the Humean doctrine himself, observes its awkwardness for the friends of facts in 'Truthmaking and Difference-Making', pp. 611–12.

devoid of necessary connections. Of course, we can allow that there is no necessary connection between the constituents of the fact in question, since both Mont Blanc and the universal property of being snow-capped can, in some possible worlds, fail to stand to one another in the exemplification relation and exist instead as constituents of other, quite different facts. However, in any ontology which includes facts, it is difficult to see how any object or universal could fail to be a constituent of *some* fact and, moreover, it seems that there will be necessary connections between distinct facts. Hence, such an ontology cannot plausibly represent all existing entities as being, as Hume would put it, 'entirely loose and separate'.

11.6 THE PRINCIPLE OF NON-CONTRADICTION AND THE INDIVISIBILITY OF TRUTH

It appears, then, that we are being driven towards a theory of truth which requires truthmakers to belong to many different ontological categories and to stand in various kinds of necessary connection to one another. The categorial diversity of truthmakers may appear to be a disunifying feature, even if the presence of necessary connections to some extent compensates for this. Certainly, we seem no nearer to establishing any interesting sense in which 'truth is one'. We have concluded that truth-*bearers* are many, contrary to the opinion of absolute idealists, who appear to hold that, strictly speaking, there is at most only one truth, which is the whole truth and nothing but the truth, and that what we commonly regard as distinct and separate truths are at best only 'partial' truths.[12] We have also concluded that truth*makers* are many, in opposition to those who hold that the True, or the world as a whole, is what makes all true propositions true. And now we have also concluded that the many truthmakers do not even belong to the same ontological category, but can be entities as different in nature as objects and properties—to which one might add, no doubt, relations, events, and perhaps even absences or privations. But despite these pluralistic features of truth, still there is a fundamental sense in which truth really is one and indivisible—a sense which is intimately connected with metaphysical realism and the oneness of the world, as I shall now try to show.

Perhaps the most fundamental logical principle governing truth is *the principle of non-contradiction*, according to which a proposition cannot be both true and false—or, as it might more strictly be put, a proposition and its negation cannot both be true. Of course, so-called 'dialetheists' challenge this principle, often in order to provide solutions to the logical paradoxes, such as the Liar and

---

[12] As F. H. Bradley puts it, '[T]ruth itself would not be complete, until it took in and included all aspects of the universe': see *Appearance and Reality* (London: Swan Sonnenschein, 1893), pp. 546–7.

Strengthened Liar paradoxes.[13] I shall not consider that position here, because I am far from being convinced that dialetheism provides the best, much less the only, way of handling these paradoxes. But why *shouldn't* a proposition be both true and false? In order for a proposition (or, at least, a contingent non-negative proposition) to be true, it must have a truthmaker, if our preceding contentions were correct. And in order for that same proposition also to be false, presumably, either it must lack a truthmaker or else its negation must have a truthmaker. Suppose, following an earlier suggestion, we adopt the first of these two alternatives and contend that a false proposition is one that lacks a truthmaker (on the assumption that it is the sort of proposition that would need to have a truthmaker in order to be true, which is the case, we may suppose, with all contingent non-negative propositions). Well, then, why shouldn't a proposition both possess a truthmaker and also lack a truthmaker? The answer seems obvious enough: its possessing a truthmaker just *is* its not lacking a truthmaker. 'Possessing' *means* 'not lacking'. But we may press our questioning further, perhaps, and ask: why shouldn't a proposition both *not lack* (that is, possess) a truthmaker and also *lack* a truthmaker? Here we may be tempted to appeal once more to the principle of non-contradiction and reply that, since the proposition that a certain proposition, *p*, does not lack a truthmaker is the *negation* of the proposition that *p* lacks a truthmaker, and the principle of non-contradiction does not permit both of these propositions to be true, we may conclude, indeed, that *p* cannot both not lack a truthmaker and lack a truthmaker. But, of course, this is just arguing in a circle, since what we were originally seeking was a reason for endorsing the principle of non-contradiction. I don't think that a non-circular argument for the principle of non-contradiction can be given, not least because an argument is a process of reasoning and as such will rely, either explicitly or implicitly, upon that very principle, so fundamental is it to the very conception of what reasoning is (*pace* the dialetheists). But this isn't to say that the principle is just a dogma which we could just as easily abandon as accept.

How *could* we abandon the principle of non-contradiction? I have mentioned 'dialetheism', but even this view doesn't contemplate the wholesale admission of contradictions, designed as it is to tackle the esoteric logical problems posed by such paradoxes as the Liar. The only way in which the principle could be abandoned wholesale, I think, is by accepting some fairly radical form of *relativism*. A relativist who contends that a proposition may be true 'for me' or 'for my culture' while also being false 'for you' or 'for another culture' is, in a reasonably clear sense, abandoning the principle of non-contradiction wholesale. Of course, such a relativist may continue to accept the principle as restricted to different believers or cultures, so that a proposition could not be true *for me* say, while also being false *for me*. Indeed, to deny even this would be to abandon intelligibility

---

[13] See Graham Priest, *In Contradiction: A Study of the Transconsistent* (The Hague: Martinus Nijhoff, 1987).

altogether, I think. But what we need now to consider are the ontological commitments of the sort of relativism that we have, so far, not ruled out as being unintelligible.

11.7 THE ONTOLOGICAL IMPLICATIONS OF RELATIVISM

If a proposition is to be true for me, say, but false for someone else, then what are we to say about it in terms of *truthmaking*? If the proposition is true for me, it will need to have a truthmaker which makes it true for me, if our preceding reasonings about truth and truthmaking have been along the right lines. And, of course, if the proposition is at the same time false for someone else, the truthmaker which makes it true for me will not be able to make it true for that other person. But how could that be? After all, if the truthmaker *exists* and the proposition is the same in both cases, how could it fail to make the proposition true for both of us, that is, true *simpliciter*? The relativist had better not answer this question by denying, after all, that it is the very same proposition that is true for me but false for the other person, for this is not any interesting sort of relativism at all: it is just a relativism of belief *contents*, whereby different believers have beliefs about different things. I think that what the genuine relativist must say at this point is that *existence itself* is relative to believers. He must say this because, I think, he must answer the question that has just been posed by saying that the truthmaker which exists to make the proposition true for me but not for you *only exists for me and not for you*. *That* is how the very same proposition can be false for you: because no truthmaker of the proposition exists *for you*.

If this is correct, we see how radical are the ontological implications of relativism. The relativist must hold that reality itself is many, not one—that we do not all inhabit the same world. He must say that the sum total of existence for me is not necessarily the same as the sum total of existence for you. This may still allow him to say that our worlds 'overlap', at least partially. After all, if he couldn't say this, then he couldn't even say that 'the other' *exists* to have beliefs which differ in truth-value from his own despite having the same contents. Moreover, if I am right about the ontological commitments of the relativist, then a further conclusion apparently follows, namely, that all existence is, in a perfectly good sense, 'mind-dependent'. All existence is mind-dependent for the relativist because what exists only exists relative to a thinker or collection of thinkers. It may not be that, for the relativist, whatever exists relative to a thinker or collection of thinkers must actually be *thought of* by one or more of those thinkers, so the relativist need not be committed to the mind-dependence of all existence in the strongest possible sense. Even so, the mind-dependence to which relativism is committed does appear to be very strong and quite at odds with any serious form of metaphysical realism.

Can we make sense of the thought that what exists for one thinker might not be the same as what exists for another? *Perhaps* we can, although this is not to say that I think that we should for one moment contemplate endorsing it. In fact, one might suppose that certain conceptions of possible worlds precisely embody a thought very much like this. One might, for instance, hold a view of possible worlds according to which individuals are 'world-bound'—in other words, according to which no individual can exist in more than one possible world—and according to which the sum total of existence is world-relative. Certain views about time are somewhat analogous, such as the view that the sum total of existence at one time differs from the sum total of existence at another. All of these views are pluralist ontologies in a very deep sense. Against them is posed a monistic ontology which holds that reality is fundamentally one: that there is just one sum total of existence—one world—which is the same for all thinkers, places and times. And my suggestion is that, to the extent that we are committed to the unity of truth, at least inasmuch as this amounts to an unconditional acceptance of the principle of non-contradiction, we are committed to the oneness of reality and to its mind-independence. We are committed, in short, to a fully realist metaphysics. Fortunately, this still leaves plenty of scope for many forms of pluralism. In accepting that reality is one, we need not accept that there is only one truth, or only one truthmaker, or only one kind of truthmaker. Our ontology will admit of multitudes within The One.

12

Truthmaking, Necessity, and Essential Dependence

12.1 TRUTH AND TRUTHMAKING

The idea that all truths need to be *made* true is an appealing one. This is so whatever one may think the 'primary' truthbearers to be—sentences, statements, beliefs, or propositions. To avoid undue complexity, I shall assume that *propositions* are the primary truthbearers in what follows, but I don't think that this assumption is crucial to the general thrust of the arguments that I shall be advancing. So, why should I say that it is an appealing idea that all true propositions need to be made true? Note here that I don't, at this stage, say that they need to be made true *by something*, and in this sense need to have *truthmakers*, construed as being existing entities of some sort—although I shall be defending such a position in due course. Well, there is plainly a difference between a proposition's being true and that same proposition's being false—and this is a difference that we obviously want to be able to explain. It doesn't follow that this difference is a difference between a proposition's possessing one property—the property of being true—and its possessing another property, the property of being false. For we can't assume without argument that the predicates 'is true' and 'is false' express properties of the entities to which they are applied. Indeed, although I am no nominalist and believe in the existence of properties—both conceived as universals and conceived as particulars that are instances of those universals—I do not consider that truth and falsity are properties in that sense. More precisely, I do not consider that a true proposition is one that exemplifies the universal *truth*, nor that it possesses a truth *trope* or *mode*, whether or not conceived as a particular instance of such a universal.

If a proposition's being true were indeed a matter of its exemplifying truth or possessing a truth trope, then I think it would not, after all, be clear why a true proposition would need to be *made* true in the sense in which it intuitively needs to be. For its being true would then just be a matter of how that proposition was 'in itself'—assuming, at least, that truth so-conceived would be a non-relational property, as the syntax of truth-predication suggests. That is to say, a proposition's being true would, on this way of conceiving the matter, be analogous to an apple's being red, or its being round. If an apple is round, it is so because it exemplifies roundness, or possesses a roundness trope. But it does not need to be *made* round, in anything like the sense in which, intuitively, a true proposition needs to be *made* true. Of course, there will need to be a *cause* of the apple's roundness and so it will need to be 'made round' in *that* sense. But it is surely no part of the intuitive idea of truthmaking that making true is a kind of causing. Rather, when we say that a true proposition needs to be 'made' true, we mean that it has to be true 'in virtue of' something, where 'in virtue of' expresses what I would call a relationship of metaphysical explanation. In other words, a true proposition must have truth conferred upon it in some way which explains how it gets to be true.

But may it not now be interjected that, likewise, a round apple needs to have roundness conferred upon it in some way which explains how it gets to be round? And couldn't it then be said that a round apple is round precisely in virtue of exemplifying the universal *roundness*, or in virtue of possessing a roundness trope? And then isn't this precisely a matter of its being 'made round', in a non-causal sense analogous to that of a proposition's being 'made true'? I think not. I happily acknowledge that a round apple is, in a perfectly good sense, round *in virtue of* exemplifying the universal *roundness* or *in virtue of* possessing a roundness trope. But this is just to say that *what it is* for a round apple to be round is for it to exemplify the universal *roundness* or to possess a roundness trope. It is not at all to say that roundness is *conferred upon it* in either of these ways. For the roundness of a round apple, we are now supposing, simply *is* either the universal *roundness* or a roundness trope—and it would just be circular to say that its roundness in either sense is *conferred* upon the apple by its exemplifying or possessing that very roundness. The only sense in which, as I put it earlier, a round apple 'needs to have roundness conferred upon it in some way which explains how it gets to be round'—the only sense in which it needs to be 'made' round—is the *causal* sense. To 'make it round', in this sense, is just to bring it about that the apple exemplifies the universal *roundness* or possesses a roundness trope. And we have already rejected the idea that a proposition needs to be 'made' true in this sense. So comparison with the case of an apple's being round in fact helps to undermine, rather than support, the suggestion that the truth predicate 'is true' expresses a universal, or that there are truth tropes that certain propositions possess.

12.2 FORMAL ONTOLOGICAL PREDICATES

The truth predicate, I consider, is best seen as belonging in the same category of expressions as such predicates as 'exists' and 'is identical with'. They are *formal ontological predicates* and we can say, if we like, that they express formal ontological properties and relations—truth, existence, and identity—provided that we don't make the mistake of supposing that such 'properties and relations' are *elements of being*, that is, existing entities in either of the ontological categories of *universal* or *trope* (see Chapter 3 above). Formal ontological properties and relations are not elements of being, to be included amongst the overall inventory of 'what there is': rather, they contribute to the nature of reality as a whole solely by helping to constitute *how* reality is. It might be thought, of course, that this was precisely the role of universals and tropes, but that would be wrong—for, according to the sort of realist position that I am defending, these *are* elements of being to be included in the overall inventory of what there is, along with other categories of entity, such as so-called concrete particulars or, to use an older terminology, individual substances. The nominalist will no doubt want to take issue with me here and contend that we no more need to regard roundness and redness as 'elements of being', whether as universals or as 'abstract' particulars, than I am saying that we need to regard truth, existence and identity in this light. The nominalist will urge that, since I have acknowledged this in the case of what I am calling formal ontological properties and relations, it is gratuitous of me to refuse to acknowledge it in the case of more mundane 'properties and relations', such as redness, roundness and betweenness. I will be accused of drawing an arbitrary line and of having started out upon a slippery slope down which I cannot, in any principled way, help sliding into full-blown nominalism.

Well, I reject the charge. A principled and non-arbitrary line *can* be drawn between those predicates that are candidates for expressing real universals and those that are not. Note, I am not implying here that every predicate that *is* a candidate for expressing a real universal must be taken actually to express one. I am perfectly happy with the contention, favoured by philosophers such as David Armstrong, that it is a largely or perhaps even wholly empirical matter which of these predicates we should regard as actually expressing real universals and, indeed, a largely or perhaps even wholly empirical matter what real universals we should suppose reality as a whole to include. All that I am saying is that it is not an empirical matter, but rather an *a priori* one, that certain predicates, such as 'is true', 'exists', and 'is identical with', are *not* candidates for expressing real universals. That is to say, I hold that we can know, purely by reflecting on the matter, that someone could not have grasped properly the meaning of such predicates if he thought that their semantic role was to express certain real universals, conceived as elements of being. No doubt the nominalist will want to reply that a similar thing could be said about *all* predicates—that, in effect, a realist construal of their semantic role, if taken seriously, could only be taken to reflect an imperfect grasp of their meaning. But I believe that any such charge is certainly open to rebuttal, because I believe that a perfectly coherent account of predication can be supplied by the realist. I don't, of course, believe that *all* predication can be understood on this model, because I don't believe that all meaningful predicates should be taken to express real universals.

So why do I 'draw the line' in the place I claim to, in distinguishing between those predicates that are candidates for expressing real universals and those that are not? Briefly, my guiding line of thought on this matter is as follows. Reality as a whole must contain what I am calling 'elements of being'. Moreover, I think that it must contain a *plurality* of such elements. Although reality is one, it is a one that embraces many. Some mystic philosophers have denied this, of course, holding that reality is one in an absolutely simple and undifferentiated way, but I can make no sense of this. For, in the sense of 'reality' now at issue, even 'appearance' is to be understood as being included in 'reality'—and appearance is certainly not, *to all appearance*, absolutely simple and undifferentiated. Moreover, I cannot make any sense of the view that reality can be cleanly divided into two domains, the domain of appearance and everything else, such that only the domain of appearance contains multiplicity and differentiation of any kind—the view, roughly speaking, that there is on the one side differentiated appearance and on the other side 'the world', with the latter being completely undifferentiated 'stuff', or 'the noumenon'. Apart from anything else, I cannot see how reality as a whole can be coherently divided into 'appearance' and 'the world', nor in any other way analogous to this, such as into 'mind' and 'world', or into 'representation' and 'reality' ('reality' in the latter case being taken, of course, to be less than what I have been calling 'reality as a whole').

12.3 ONTOLOGY, CATEGORIES, AND METAPHYSICAL REALISM

I am, thus, an ontological *pluralist*, but not an ontological *relativist*: I hold that there is just one reality, but that it embraces a multiplicity of elements of being. One of the tasks of ontology, then, is to provide an inventory of those elements of being. However, such an inventory could not intelligibly be nothing more than a gigantic washing list. It would miss the point of ontology altogether to suppose that its task were simply to enumerate all the entities that there putatively are: shoes and ships and sealing wax, cabbages and kings—*et cetera, et cetera*. Many nominalists, it seems to me, implicitly suppose that, in the last analysis, this *is* all that we may hope to do by way of characterizing the elements of being. They fundamentally agree with Quine when he said that the basic question of ontology is 'What is there?', but that it could be answered by the one-word English sentence 'Everything'.[1] He meant, I think—and meant seriously—that this is the best *general* answer that we can give to that question.

To this it may be replied that Quine himself, if not his followers, also held that the best answer to the question of what there is is to be found by determining what it is that the bound variables of our best-supported scientific theories should be taken to quantify over. And it may be added that it seems that he himself thought that the answer might well turn out to be that all that exist are *numbers and sets*, since all of our best-supported scientific theories can be interpreted most economically as

---

[1] See W. V. Quine, 'On What There Is', in his *From a Logical Point of View*, 2nd edn (Cambridge, MA: Harvard University Press, 1961).

quantifying only over such entities, rather than, say, over spatiotemporally located material objects or materially filled regions of spacetime—since any sentence putatively quantifying over entities of the latter sorts can be reinterpreted, consistently with the empirical evidence in support of the theory containing it, as a sentence quantifying over sets of numbers (to wit, numbers specifying the coordinates of what were previously conceived to be spacetime locations).[2] However, it would not be consistent with the spirit of Quine's philosophy more generally to suppose that he was seriously committed to Pythagoreanism—the view that, in reality, all that exist are mathematical objects. For Quine espoused the doctrine of *ontological relativity*, which excludes the idea that there is any privileged way of specifying the contents of reality that is wholly independent of one's means of representing it in language.[3] And what language we use is a contingent matter, determined by cultural and psychological factors. This applies quite as much to the language of science as to any other language. It may perhaps be that the language of fundamental physics and the theories expressed in that language can be interpreted, without affecting their empirical content or predictive utility, as quantifying solely over numbers and sets, and so as incurring an ontological commitment only to mathematical objects. But for a Quinean naturalist this fact should not be construed as providing support for any serious endorsement of Pythagoreanism as a contribution to fundamental metaphysics, that is, as a putative account of what the real elements of being are.

However, my concern is not to defend Quine nor to try to render his ontological pronouncements more palatable than they might seem to be. I think, in fact, that he ultimately has nothing intelligible to say about ontology and would happily accept that charge, on the grounds that ontology as I conceive of it is an impossible enterprise. By my way of thinking, however, the reason why Quine has nothing intelligible to say about ontology as I conceive of it is that he really does believe that the only perfectly general answer, of the sort that I would acknowledge as being relevant, that can be given to the question 'What is there?' is the one-word reply 'Everything'. On Quine's view, the most that the doctrine of ontological relativity will allow us to say about the nature of reality 'as it is in itself' is that *there are things*, where the term 'thing' is perfectly neutral, denoting no more than a possible value of a variable of quantification. Anything whatever could, of course, be such a value. There is no room, in Quine's view, for *categorial* differentiations amongst 'things' in any ontologically serious sense. At most he allows that different predicates may be true of different things: perhaps some things are 'cabbages', for instance, while others are 'kings'. But if we were to say that some things are 'universals', say, while others are 'particulars', he would take this as conveying no more than just another putative difference between the predicates true of different things.

---

[2] See W. V. Quine, 'Things and their Place in Theories', in his *Theories and Things* (Cambridge, MA: Harvard University Press, 1981).

[3] See W. V. Quine, 'Ontological Relativity', in his *Ontological Relativity and Other Essays* (New York: Columbia University Press, 1968).

It may be instructive to compare Quine's minimalist ontology with various so-called 'one-category' ontologies. One such ontology is the pure trope ontology, according to which the only elements of being are tropes or property-instances.[4] Another is the ontology of classical resemblance nominalism, according to which the only elements of being are concrete particulars, or individual substances.[5] Ontologists espousing these views certainly do not think—or ought not to think—that belonging to an ontological category is *just* a matter of being describable by a predicate: that 'is a trope' or 'is a concrete particular' is just on a par with 'is a cabbage' or 'is a king', the former predicates differing from the latter merely in that they are universally applicable. If they thought that, it would simply be unintelligible that there was any dispute between them as to which view was correct. Neither of them thinks that telling whether something is a trope as opposed to a concrete particular is remotely like telling whether something is a cabbage as opposed to a king. The trope theorist holds that what we call cabbages are in fact 'bundles' of tropes, whereas the classical resemblance nominalist holds that they are entities—concrete particulars as he conceives of them—which admit of no decomposition into anything other than further concrete particulars. Their difference is a difference concerning the ontological status of the entities to which *all* of our descriptive predicates apply. Two such ontologists could agree perfectly about how to *describe* the world—agree, for instance, that it includes shoes and ships and sealing wax, cabbages and kings, along with anything else that one could expect to find on a gigantic 'washing list'. They differ over the nature of entities to which these descriptions apply—whether or not cabbages, for instance, are 'bundles of tropes'. But Quine has no serious interest in any such dispute. His is most aptly characterized not as a *one*-category ontology—the one category being 'thing' in the broadest possible sense, or 'entity'—but rather as a *no* category ontology. On Quine's view, all that we can ever do is to disagree about how to describe what there is, not over what there is to be described.

But a no category ontology is an incoherent ontology. For either it maintains that what there is is many, or that what there is is one, or that what there is is neither many nor one. As I have already explained, my own view is that the only coherent position is that although reality is one, it contains multiplicity—so that what there is is many. Quine himself *seems* to suppose so too, for he holds that 'to be is to be the value of a variable' and seems to be committed to the multiplicity of such values. Pythagoreanism would certainly respect the principle that what there is is many—many mathematical objects, including all the numbers. But Quine also espouses the dictum 'No entity without identity' and in some sense that must be correct too.[6] For how can there be multiplicity where there is neither identity

---

[4] See Keith Campbell, *Abstract Particulars* (Oxford: Blackwell, 1990).
[5] See Gonzalo Rodriguez-Pereyra, *Resemblance Nominalism: A Solution to the Problem of Universals* (Oxford: Clarendon Press, 2002).
[6] See, for example, W. V. Quine, 'Speaking of Objects', in his *Ontological Relativity and Other Essays*.

nor distinctness? There can only be many if each of the many is a one that is identical only with itself and distinct from each of the rest. However, a no category ontology leaves no scope for any real difference between one and many nor between identity and distinctness. Given his no category ontology, Quine's one-word answer to the question 'What is there?'—'Everything'—is misleading to the extent that it suggests that what there is is determinately and objectively either one or many. For the Quinean, all questions concerning 'how many' things there are and which things are identical with or distinct from one another have to do with how we describe reality, not with what reality contains prior to or independently of our attempts to describe it. Thus Quine is implicitly quite as committed to the 'amorphous lump' conception of reality as Michael Dummett is explicitly committed to it.[7] Both of them are anti-realist metaphysicians in the fullest sense of the term, because the distinction between an utterly formless 'something' and nothing at all is a distinction without a meaningful difference. Indeed, in the end they are both *nihilist* metaphysicians, because there is no coherent way for them to exempt *us* and our *descriptions* or *thoughts* about reality from the annihilating acid of their anti-realism.

12.4 IDENTITY, ESSENCE, AND ESSENTIAL DEPENDENCE

The preceding discussion may seem to have taken us far from the topic of truthmaking, but the digression has not been an irrelevant one. Its purpose was to defend the view that formal ontological predicates need to be understood in a different way from ordinary, empirical or descriptive predicates. These formal ontological predicates include those used to assign entities to certain ontological categories, such as the predicates 'is a trope' and 'is a concrete particular'. They also include, I maintain, such predicates as 'is true', 'exists' and 'is identical with'. We should not expect the basis on which these predicates are correctly applied to entities to be at all similar to the basis on which empirical or descriptive predicates are correctly applied to them. It is part of the task of metaphysics to explain how empirical or descriptive predicates may be correctly applied to entities by appealing to formal ontological features of and relationships between those entities. For example, a trope theorist may explain how the descriptive predicate 'is red' is correctly applicable to an entity by saying that the entity in question is a bundle of tropes which includes a trope belonging to a certain resemblance class of tropes. A classical resemblance nominalist may explain the same thing by saying that the entity in question is a concrete particular which itself belongs to a certain resemblance class of concrete particulars. Neither of them, of course, would—or

---

[7] See Michael Dummett, *Frege: Philosophy of Language*, 2nd edn (London: Duckworth, 1981), pp. 563 ff.

coherently could—apply the same explanatory strategy to explain how the predicates 'is a trope' or 'is a concrete particular' are correctly applicable to entities.

How, then, *are* we to explain how formal ontological predicates are correctly applicable to entities? In some cases, I want to say, such a predicate applies to an entity *in virtue of its identity*, that is, in virtue of *what that entity essentially is*. It is for this reason that the predicate 'is a trope' or the predicate 'is a universal' applies to any entity. The same reason obtains in the case of the identity predicate itself, 'is identical with'. This predicate is correctly applicable to an entity x and an entity y in virtue of the identity of x and the identity of y—and, of course, if it is correctly applicable to 'them', then 'they' are one and the same entity, with the same identity. But, equally, the distinctness predicate, 'is distinct from', is correctly applicable to entities in virtue of their identities. However, at least some entities, I want to say, *depend* for their identities on the identities of other entities. What this means, as I understand it, is that it is *part of the essence* of such an entity that it is *the very entity that it is* in virtue of a unique relationship in which it stands to one or more other entities. (I use the term 'essence' here in precisely the way that Locke recommended, to denote 'the very being of any thing, whereby it is, what it is'—which, he says, is the 'proper original signification' of the word.[8]) This is my own view about, for instance, tropes—or, as I prefer to call them, *modes*. I hold that if m is a mode or trope—suppose, for example, that m is a certain *roundness* mode—then it depends for its identity on the identity of the concrete particular or individual substance that possesses it—a certain apple, say. This is because, in my view, it is part of the essence of m that it is *the very entity that it is*—this roundness mode as opposed to any other exactly resembling roundness mode—in virtue of being the roundness mode *that is possessed by this apple*. Pure trope theorists, of course, cannot take the same view of what *they* call 'tropes', since they do not believe in the existence of either universals or individual substances. So although both my 'modes' and the pure trope theorist's 'tropes' may loosely be termed 'property-instances', they are in fact entities belonging to rival and quite distinct ontological categories, because the entities belonging to those putative categories are quite different in respect of 'what they essentially are', that is, in respect of their 'identities'. The pure trope theorist must apparently hold that each trope has its identity underivatively, not that it *depends for it on* or *owes it to* other entities of any sort.

Identity dependence in the foregoing sense is a species of *essential dependence*.[9] But there are other species as well. Very plausibly, an entity can, for example,

---

[8] See John Locke, *An Essay Concerning Human Understanding*, ed. P. H. Nidditch (Oxford: Clarendon Press, 1975), Book III, ch. III, sect. 15.

[9] For more discussion and a definition of identity dependence, see my *The Possibility of Metaphysics: Substance, Identity, and Time* (Oxford: Clarendon Press, 1998), ch. 6. My thoughts about these matters have been helped by some of Kit Fine's work on the subject, but nothing that I say concerning them should be assumed to coincide with his own views. See, in particular, his 'Essence and Modality', in James E. Tomberlin (ed.), *Philosophical Perspectives, 8: Logic and Language* (Atascadero, CA: Ridgeview, 1994) and 'Ontological Dependence', *Proceedings of the Aristotelian Society* 95 (1995), pp. 269–90.

depend essentially for its *existence* on one or more other entities, without necessarily depending essentially for its *identity* upon those other entities. Immanent universals seem to provide a case in point. Observe that, very plausibly, it is *not* part of the essence of the universal *roundness* that it has the roundness of a certain apple as one of its instances, for the simple reason that the universal could have lacked that particular instance. This is because the apple is a contingent being and, moreover, exemplifies roundness only contingently—it could have lacked the roundness mode that it actually possesses. Consequently, it seems, the universal does not depend for its *identity* upon this or indeed—I should say—any other mode that is an instance of it. Even so, if the universal *roundness* is an *immanent* universal—as I hold it and all other universals to be—then it seems that it does, in a perfectly good sense, *depend essentially for its actual existence* on the roundness modes of all actually existing round individual substances. For, on the immanent conception of such a universal, it is part of the essence of the universal that it actually exists only if it is actually exemplified by certain individual substances, in virtue of those substances possessing modes that are particular instances of the universal. If *other* substances had exemplified the universal, then the universal would have depended for its existence on the modes of the universal that would in *that* case have existed—but, again, it seems that it would have depended upon them *essentially*, again because it is part of the essence of an immanent universal to exist only if it has particular instances.

12.5 THE VARIETIES OF METAPHYSICAL DEPENDENCE

It is important to recognize that essential dependence is not the only variety of metaphysical dependence. Sometimes, for instance, an entity can, in a perfectly good sense, depend metaphysically for its existence on another entity, even though it does not depend *essentially* for its existence on that entity. Suppose, for example, that mathematical objects such as the natural numbers exist and are necessary beings—that is to say, that they are beings 'whose essence includes existence'. What this means is that it is, supposedly, part of the essence of a number, such as the number 7, that it exists. One might suppose, indeed, that the number 7 does not depend essentially for its existence on anything else, not even God or other numbers—although in fact it is, I think, plausible to say that it depends essentially for its existence at least on other numbers, for it is plausibly part of the essence of the number 7 that it stands in certain arithmetical relations to other numbers. Be that as it may, consider now a contingent being, such as a certain apple. This apple plausibly depends essentially for its existence, both actual and possible, upon certain other contingent entities. For instance, it is plausibly part of the essence of this apple that it actually exists only because a certain apple tree actually exists—the apple tree on which it grew—so that the apple depends essentially for its actual existence on that apple tree. However, it is *not* plausibly part of the essence of this apple that it actually exists only because the number 7 exists. For it is surely no part of what this apple essentially *is* that is related to an abstract mathematical object like the number 7 in any way. None the less, it is clearly the case that this apple *could not have existed* without the number 7 existing, simply because—or so we are assuming—the number 7 is a necessary being. But this is just to say that this apple stands in a certain relation of metaphysical dependence to the number 7: the relation in which one thing stands to another when it is metaphysically necessary that the first thing exists only if the second thing does. We can call this species of metaphysical dependence 'necessary dependence', to distinguish it from what we have been calling 'essential dependence'. In Chapter 3 above, I called it 'rigid' existential dependence.

We have just seen an example of something that depends *necessarily* for its existence on a certain other thing, even though it does not depend *essentially* for its existence on that other thing—the things in question being, in this case, a particular apple and the number 7. But it seems that it can also be the case that something depends *essentially* for its existence—at least, for its *actual* existence—on some other thing or things, even though it does not depend *necessarily* for its existence (in the sense just defined) on the other thing or things in question. In effect, we saw an example of this earlier, in the case of the immanent universal *roundness* and the roundness modes of all actually existing round substances. For I said that the universal *roundness* depends essentially for its *actual* existence upon those roundness modes—indeed, that this is what it is for such a universal to be 'immanent'. And yet it is clearly *not* the case, as I pointed out earlier, that it is metaphysically necessary that the universal *roundness* exists only if those roundness modes exist, for that very universal clearly could have existed even if none of those roundness modes had existed, provided that other roundness modes had existed instead.

Notice, incidentally, that a universal's *essential* dependence for its actual existence upon its actual modes cannot simply be identified with what, in Chapter 3, I called its *non-rigid* existential dependence upon those modes, which is clearly a *weaker* kind of metaphysical dependence. Indeed, the problem that we encountered there in defining non-rigid existential dependence in a way which avoids triviality does not, it seems, arise for the notion of *essential* existential dependence. One lesson of this is that it is preferable to appeal, as I have done in the present chapter, to the latter notion in order to specify the distinctive manner in which immanent universals depend for their existence upon their particular instances. However, this does not, I believe, significantly affect conclusions that I have reached elsewhere in this book in arguing for which I have relied instead upon the weaker notion of non-rigid existential dependence.

12.6 TRUTHMAKING AS ESSENTIAL DEPENDENCE

We are now in a position to apply some of these considerations to the case of *truth*. I said at the outset that it is an appealing idea that all truths need to be *made* true. We can now try to cash out this idea in terms of the notion that the truth of any proposition is a metaphysically dependent feature of it—remembering that by 'feature' here I do not mean a *property*, either in the sense of a universal or in the sense of a mode or trope. The idea, then, is that any true proposition depends metaphysically for its truth on something. Now, conceivably, a proposition might depend metaphysically for its truth simply on *itself*. This might be the case, for instance, with logically necessary truths. But most propositions are surely not like that. In any case, in order to proceed further, we need to consider what *species* of metaphysical dependence is most plausibly involved in truthmaking. I think it is most plausibly *essential* dependence that is involved, rather than *necessary* dependence. We can begin to see why by noting that, analogously with the case of existence, we should say that a proposition depends necessarily for its *truth* on a certain entity just in case it is metaphysically necessary that the proposition is true only if the entity in question exists. However, this means, for instance, that the proposition that this apple is round depends necessarily for its truth on *the number 7*—for it is, clearly, metaphysically necessary that the proposition that this apple is round is true only if the number 7 exists, simply because (as we are supposing) the number 7 is a necessary being. But it would surely be quite inappropriate to say that the number 7 is a 'truthmaker' of the proposition that this apple is round.

However, it may be suggested that what is wrong with the foregoing proposal is not that it invokes the notion of metaphysical necessity, but just that it invokes it in the wrong way. The reason why the number 7 cannot be a truthmaker of the proposition that this apple is round, it may be said, is just that that proposition could have been *false* despite the existence of the number 7 and a truthmaker is not, or not merely, something whose existence is necessary for the truth of a proposition but something whose *non*-existence is necessary for its *falsehood*—in other words, it is something whose existence *metaphysically necessitates* the truth of the proposition. In the language of possible worlds, it is something that exists not, or not merely, in every possible world in which the proposition in question is true, but in no possible world in which it is false. However, this view of truthmaking presents some serious difficulties. First, it is now being implied that if something is a truthmaker of a certain proposition, then it is something such that it is metaphysically necessary that this thing exists only if that proposition is true—and hence that the existence of a truthmaker is, in that sense, metaphysically dependent on the truth of any proposition that it supposedly 'makes true'. But this seems to reverse the proper direction of dependence between truthmakers and truth. Secondly, a difficulty arises regarding the truthmakers of necessary truths.

For, because any necessary being exists in every possible world and any necessary truth is false in no possible world, it turns out, according to the proposal now under consideration, that any necessary being is a truthmaker of any necessary truth, so that the truthmaking relation becomes utterly indiscriminate where necessary truths are concerned.

What we should say, I believe, in order to avoid these difficulties is that truthmaking involves a variety of *essential* dependence. A truthmaker of a proposition, I am inclined to say, is something such that *it is part of the essence of that proposition that it is true if that thing exists*. (Notice that I just say 'if' here, not 'if and only if', for reasons that will become plain in due course; we shall also see later on that there is a certain difficulty attending the proposal that I have just advanced, so that I think that it can only be regarded as a first approximation to what is wanted.) This account of truthmaking enables us to say, as was suggested earlier, that any proposition that is a logically necessary truth is *its own* truthmaker: for, plausibly, it is indeed part of the essence of, say, the proposition that *nothing both is and is not*—the law of non-contradiction—that it is true if it exists. And since, plausibly, this proposition is also a necessary being, in whose essence it is to exist, it follows that it is part of the essence of the proposition that nothing both is and is not that it is *unconditionally* true, as befits a law of logic. (So perhaps the chief difference between purely logical truths and other propositions is that, while it is part of the essence of *any* proposition that it is *either true or false*, only in the case of a proposition that is a purely logical truth is it part of the essence of that proposition that it is *true*.) Metaphysically necessary truths that are *not* logically necessary should not be seen in the same light: they are *not* their own truthmakers. Consider, for example, the mathematically necessary truth that *7 plus 5 equals 12*. In this case, it seems that the proper thing to say is that it is part of the essence of this proposition that it is true if the natural numbers exist—or, at least, if the numbers 5, 7 and 12 exist. *These numbers* are truthmakers of the proposition in question. For it is upon these numbers that the truth of that proposition essentially depends, because it is part of the essence of these numbers that they stand in the relevant arithmetical relation. So we can already see that by appealing to the notion of essential dependence to explain the idea of truthmaking, we can avoid the unwanted implication that the truthmaking relation is utterly indiscriminate where necessary truths are concerned—for we have already been able to discriminate between the truthmakers of logically and arithmetically necessary truths.

To illustrate this point further, suppose that it is a metaphysically necessary truth that *God is omniscient*, on the grounds that it is part of God's essence that he is omniscient. Then we can say that it is, equally, part of the essence of the proposition that God is omniscient that it is true if God exists—and hence that God himself is a truthmaker of this proposition. However, it would seem that he is not its only truthmaker, *God's omniscience* being another—where by 'God's omniscience' I mean the particular omniscience of God, which is a mode (and, perhaps, necessarily the *only* mode) of the universal *omniscience*. And it would certainly seem to be part of the essence of the proposition that God is omniscient not only that it is true if God himself exists, but also that it is true if God's omniscience exists. For, after all, it is surely part of the essence of God that he exists if and only if his omniscience exists and, equally, part of the essence of God's omniscience that it exists if and only if God exists. I am aware that saying this may lay me open to a charge of heresy on some accounts, because I seem to be distinguishing between God and his omniscience in a way which might seem to challenge the doctrine of God's simplicity! However, I can proceed no further at present in these theologically deep waters, so offer the example only for the purposes of illustration without presuming that anything that I have said about it is ultimately defensible from either a metaphysical or a theological point of view.

12.7 WHY FACTS ARE NOT NEEDED AS TRUTHMAKERS

It will be noticed that I have not so far invoked *facts* as truthmakers of propositions, nor do I desire to do so. It suffices, I believe, to invoke only entities in the ontological categories of *universal, individual substance* and *mode* for these purposes. Facts are typically invoked as truthmakers by those philosophers who believe in the existence of universals and individual substances (or 'concrete particulars'), but not in the existence of modes or tropes.[10] They need them for this purpose because in the case of a contingently true predicative proposition of the form 'a is F', where 'a' denotes an individual substance and 'F' expresses a universal, Fness, neither a nor Fness nor the pair of them can be a truthmaker of the proposition in question, on any remotely acceptable account of truthmaking (see further Chapter 11 above). Consequently, these philosophers invoke a new kind of entity, a's being F, or a's exemplifying Fness, which supposedly has both a and Fness as 'constituents' but is in some way more than just the conjunction or sum of a and Fness, and take *this* to be the truthmaker of the proposition that a is F. But, to my way of thinking, this manner of proceeding is mystery-mongering to no good purpose, brought about simply because the philosophers in question have tried to do without one of the fundamental categories of being.

The mysterious element in their account emerges when we ask about the nature of the supposedly contingent 'connection' between the constituents of a supposedly contingent fact. Notoriously, it will not do to regard this as being a further 'constituent' of the fact, on pain of falling into Bradley's famous regress (see Chapter 7 above). To label the connection a 'non-relational tie' is just to give a name to the mystery without solving it. It seems that the only tenable way of

---

[10] See, especially, David M. Armstrong, *A World of States of Affairs* (Cambridge: Cambridge University Press, 1997).

proceeding is to abandon any idea that facts are somehow 'composed' of their alleged 'constituents' and hold instead that the 'constituents' of facts are mere *abstractions from*, or *invariants across*, the totality of facts, identifying that totality with what Wittgenstein called 'the world'. However, then it is obscure how any fact can really be contingent, because if the constituents of facts are just abstractions from the facts containing them, it would seem that the very identity of any such constituent must be determined by its overall pattern of recurrence in the totality of facts. That is to say, it becomes impossible to see how any given constituent could have been a constituent in a possible totality of facts different from the actual totality. But in that case, the very point of invoking facts in the first place has been undercut, since they were invoked to provide truthmakers of supposedly contingent predicative propositions—and yet now it seems that there can be no such propositions. All predicative propositions turn out to be necessary truths, because the identity of any individual substance is now taken to be determined, in quasi-Leibnizian fashion, by the totality of predicative truths concerning it.[11]

12.8 HOW ARE CONTINGENT TRUTHS POSSIBLE?

How, then, can the possibility of contingent predicative truths be preserved according to an ontology which eschews facts as their supposed truthmakers? In the following manner. Consider again the proposition that this apple is round, which would seem to be a contingent one if any proposition is. According to an ontology which includes not only individual substances, such as this apple, and universals, such as the universal *roundness*, but also *modes*, such as this apple's roundness, any roundness mode of this apple would be a truthmaker of the proposition that this apple is round. For any roundness mode of this apple depends essentially for its existence both on this apple and on the universal *roundness*. Consequently, it is part of the essence of any roundness mode of this apple that it exists only if both this apple and the universal *roundness* exist. More than that, however, it is clearly part of the essence of any roundness mode of this apple that it exists only if this apple exemplifies the universal *roundness*, that is, only if the proposition that this apple is round is true. Notice that I say 'only if', not 'if and only if'—for, clearly, it is *not* part of the essence of any roundness mode of this apple that if the proposition that this apple is round is true, then *that* roundness mode exists: for this apple could have been round in virtue of possessing a different roundness mode. Nonetheless, it seems clear that what we *can* say is that it is part of the essence of the proposition that this apple is round that it is true if and only if *some* roundness mode of this apple exists. However, it is

---

[11] This seems to be the position towards which Armstrong himself has been gravitating: see David M. Armstrong, *Truth and Truthmakers* (Cambridge: Cambridge University Press, 2004). See also Chapter 7 above.

a contingent matter whether or not any roundness mode of this apple exists, because this apple is not a necessary being and, moreover, it is not part of the essence of this apple that it exemplifies the universal *roundness*—in other words, it is not part of the essence of this apple that it possesses any roundness mode. The relation of essential dependence between this apple and any roundness mode that it may possess is *asymmetrical*: it is part of the essence of any such roundness mode that it is possessed by this apple, but it is not part of the essence of this apple that it possesses any such roundness mode.

What we have found, in effect, is the sort of thing that can provide the 'contingent link' between an individual substance, such as this apple, and a universal, such as the universal *roundness*, which has to exist in order for the proposition that this apple is round to be true. Any roundness mode of this apple can provide such a link, for any such mode is a contingent being whose existence suffices to guarantee that this apple exemplifies the universal *roundness*. For it is part of the essence of any such roundness mode that it is an instance of the universal *roundness*, and it is also part of the essence of any such roundness mode that it is possessed by this apple. Consequently, if such a mode exists, the apple possesses it and the mode is an instance of the universal *roundness*, whence it follows that the apple exemplifies the universal: it does so simply in virtue of possessing a mode that is an instance of that universal. Some philosophers who do not countenance the existence of modes make the mistake of trying to let the relationship of exemplification *itself* provide the 'contingent link' between an individual substance, *a*, and a universal, *F*ness, which has to exist (in addition to *a* and *F*ness) in order for the proposition that *a* is *F* to be true. But this strategy is bound to fail, because the exemplification 'relation' is only a formal ontological relation, not a genuine 'element of being'. Consequently, it is not something that can be said to *exist* at all. Other philosophers, acknowledging this point, nonetheless assume that the 'relation' of exemplification is a *direct* formal ontological relation between the substance *a* and the universal *F*ness. But then the problem is to see how this relationship can obtain merely *contingently*. For, given that *a* and *F*ness do not *need* to stand in this relationship to one another—given, that is, that it is not part of the essence of either of them that they do so—what could possibly explain whether or not they actually do so? Notice that a *causal* explanation is not what is being sought at this point, but a metaphysical one.

It is hopeless, really, to appeal to a notion like 'unsaturatedness' here, quite apart from the fact that it rests on a dubious metaphor. This is the idea that a universal such as *F*ness is 'incomplete' in itself, rather like a chemical ion whose outer electron shell awaits completion by the addition of one or more other electrons— and that the 'completion' can be brought about by the 'union' of *F*ness with an individual substance, such as *a*. The idea is hopeless because it remains utterly obscure how the 'union' is achieved. In the case of a chemical ion, what happens is that some free electrons are attracted into the ion's outer shell by the electrostatic force exerted by the ion's nucleus—and this force is certainly something real, an 'element of being'. Nothing analogous is available in the case of *a* and *F*ness, when exemplification is conceived of as a direct formal ontological relation between them.

Modes provide the answer to the problem. Modes are real beings which stand in non-contingent formal ontological relations both to individual substances and to immanent universals. When an individual substance possesses a certain mode, it is part of the essence of that mode that it is possessed by that substance, although not part of the essence of that substance that it possesses that mode: they stand in a relationship of asymmetrical essential dependence to one another. Similarly, it is part of the essence of any mode that it is an instance of a certain universal, although not part of the essence of that universal that it has that mode as an instance: they too stand in a relationship of asymmetrical essential dependence to one another. An individual substance *exemplifies* a given universal just in case a mode exists which stands in two such relationships of asymmetrical essential dependence, one to the substance and one to the universal. (Here I am only considering the case of 'occurrent' exemplification, as discussed in Chapter 8—but 'dispositional' exemplification can be accommodated in another way.) Thus exemplification is an *indirect* formal ontological relationship between individual substances and universals and one that can be *contingent* because it can be a contingent matter whether an appropriate mediating mode actually exists.

12.9 PROPOSITIONS AND WHAT THEY ARE 'ABOUT'

I have been arguing that the intuitively attractive idea of truthmaking is best cashed out in terms of the notion of essential dependence. More specifically, I have been maintaining that a *truthmaker* of any given proposition is something such that it is part of the essence of that proposition that it is true if that thing exists—although, as we shall shortly see, this account may need to be modified in a certain way. And by the 'essence' of any entity I mean that in virtue of which it is the very entity that it is. So, for example, it is part of the essence of any entity that it belongs to a certain ontological category and also part of its essence that it is *that* member of the category in question as opposed to any other. Thus, assuming that there are such entities as propositions and that they comprise an ontological category—even if only a sub-category of some more fundamental category—it will be part of the essence of any given proposition that it is a *proposition* and that it is *that* proposition as opposed to any other. But at least some propositions depend for their identities on entities belonging to other ontological categories. Any proposition that is, as we say, 'about' certain other entities depends for its identity on those entities. Thus, the proposition that a certain individual substance, *a*, exemplifies a certain universal, *F*ness, depends for its identity on both *a* and *F*ness: for the proposition is *essentially* 'about' those other entities and so couldn't be the very proposition that it is without being 'about' *them*. It may not necessarily follow from this that it is part of the essence of the proposition that *a* is *F* that *a* and *F*ness both *exist*. If that *were* so, it would mean that the proposition depends essentially for its *existence* on the existence of *a* and *F*ness, and so cannot exist unless they exist. However, one might be able to argue that a proposition can exist while depending for its identity on other entities which do not exist. I am by no means convinced that this *can* be argued successfully, but am prepared to keep an open mind about the matter for present purposes.

What *is* clear, however, is that—at least in certain cases—if a proposition is 'about' certain entities and hence depends for its identity on those entities, then it is part of the essence of that proposition that it is *true* only if those entities exist. I say 'at least in certain cases'—and the cases that I have centrally in mind are contingent propositions about the entities in question, where these entities are themselves contingent beings. Possible exceptions would be propositions such as 'Either *a* is *F* or *a* is not *F*', which I would not describe as being a 'pure' logical truth, because it is 'about' certain non-logical entities. Conceivably, it might be held that this proposition is true, in virtue of its logical form, even in circumstances in which *a* and *F*ness do not exist. Be that as it may, what we now have to notice about truthmaking is this. A truthmaker of a proposition need not, in general, be something upon which the identity of that proposition depends. Thus, for example, I argued earlier that the proposition that *a* is *F* has as a truthmaker any mode of *F*ness possessed by *a*. On the view of truthmaking that I have proposed, this commits me to saying that such a mode is something such that it is part of the essence of the proposition in question that it is true if that thing exists. This raises a certain difficulty that I shall discuss shortly, but let us go along with the suggestion at least for the time being. So suppose that *m* is a certain mode of *F*ness possessed by *a*. Then, I want to say, *m* is a truthmaker of the proposition that *a* is *F*—and yet the proposition that *a* is *F* by no means depends for its identity upon *m*. For the proposition is not 'about' *m* at all: it is only 'about' *a* and *F*ness. So this serves to illustrate my point that a truthmaker of a proposition need not, in general, be something upon which the identity of that proposition depends.

Incidentally, we can now see clearly why it was important to say that a truthmaker of a proposition is something such that it is part of the essence of that proposition that it is true *if*—not if *and only if*—that thing exists. For it simply isn't the case, for instance, that the proposition that *a* is *F* is true only if *m* exists: it could have been true in virtue of the existence of *another* mode of *F*ness possessed by *a*. At the same time, it is worth emphasizing once again why we need to characterize truthmaking in terms of *essential* dependence rather than merely in terms of what I earlier called *necessary* dependence: that is, why we should *not* say that a truthmaker of a proposition is something such that it is metaphysically necessary that that proposition is true if—or, indeed, if and only if—that thing exists. For, as we noted earlier, this would make any necessary being a truthmaker of any necessary truth, quite indiscriminately.

12.10 A SKETCH OF A THEORY OF TRUTHMAKING

Now, however, I need to confront the difficulty that I alluded to a moment ago. I said that if *m* is a certain mode of *F*ness possessed by *a*, then *m* is a truthmaker of the proposition that *a* is *F*—and I certainly don't want to give up that claim. But I have also been maintaining that a truthmaker of a proposition is something such that it is part of the essence of that proposition that it is true if that thing exists. This means that I must claim that it is part of the essence of the proposition that *a* is *F* that it is true if *m* exists. But this is problematic if one thinks, as one may well do, that no entity can be in any manner essentially dependent on any entity that could fail to exist in circumstances in which that entity itself did exist. It was this thought that motivated the doubt, mentioned earlier, that an entity could depend for its *identity* on something non-existent. Now, clearly, we must be able to say that the proposition that *a* is *F* could be true in circumstances in which *m* itself failed to exist: but then the constraint on essential dependence that is now being proposed would rule out our saying that it is part of the essence of the proposition in question that it is true if *that* mode of *F*ness, *m*, exists. Perhaps the most that we can say about the essence of the proposition that *a* is *F* is that it is part of the essence of that proposition that it is true if *some* mode of *F*ness exists. But then how do we define the truthmaking relation in terms of essential dependence, given the constraint being proposed, on the assumption that it is to be defined as a relation between entities—or, more precisely, as a relation between an entity and a proposition?

One possibility would be to say that an entity *e* is a truthmaker of a proposition *p* if and only if it is part of the essence of *p* that *p* is true if some entity relevantly similar to *e* exists—for example, a mode of *F*ness possessed by *a*, in the case in which *e* is such an entity. But the notion of 'relevant similarity' invoked here is much too vague for our purposes (though I shall suggest a way of rendering it more precise in a moment). An alternative possibility would be to appeal to the essence of *e* rather than to the essence of *p* and say that an entity *e* is a truthmaker of a proposition *p* if and only if it is part of the essence of *e* that *p* is true if *e* exists. But this would be implausible, because it is most implausible to say, for example, that it is part of the essence of a certain mode of *F*ness possessed by *a*, *m*, that a certain proposition 'about' *a* and *F*ness is true if *m* exists. For it is most implausible—or so it seems to me—to say that an entity such as *m* has it as part of its essence that it is related in any way whatever to any *proposition*.

What we are looking for is a satisfactory way of completing the following biconditional statement, intended as a definition of the truthmaking relation: 'An entity *e* is a truthmaker of the proposition *p* if and only if . . .'. And the problem that has just been raised is a problem for our original proposal that this should be completed by the clause 'it is part of the essence of *p* that *p* is true if *e* exists'. But it is more specifically only a problem for the right-to-left reading of the biconditional.

There is no problem in saying that *if* it is part of the essence of a proposition *p* that *p* is true if *e* exists, *then e* is a truthmaker of *p*. The problem only arises for the claim that if *e* is a truthmaker of the proposition *p*, then it is part of the essence of *p* that *p* is true if *e* exists—the problem being that this seems to require us to allow that it can be part of the essence of a proposition that it be related to an entity which might fail to exist in circumstances in which that proposition itself did exist. It is obvious that we shall be required to allow this if we hold that all propositions are necessary beings but that some of their truthmakers are not. But, as we have seen, we shall still be required to allow it if, for example, we hold that a proposition exists only in circumstances in which all of the entities that it is 'about' exist—if, for instance, we hold that the proposition that *a* is *F* exists only in circumstances in which both *a* and *F* exist. For then a certain mode of *F*ness possessed by *a*, *m*, need not exist in every circumstance in which that proposition exists, nor even in every circumstance in which that proposition is *true*.

I suspect that there is no simple and straightforward way to deal this problem. But rather than leave the issue entirely unresolved, I shall instead tentatively offer a suggestion for an alternative approach. It may be that rather than simply trying to find a way to complete the foregoing biconditional statement in order to specify the relation of truthmaking, we should proceed in a more roundabout manner, allowing a specification of the class of truthmakers to emerge out of a *theory* of truthmaking. The axioms of the theory might be taken to be something like the following. (1) For any proposition, *p*, there are one or more types of entity, $E_1, E_2, \ldots E_n$, such that, for any *i* between 1 and *n*, it is part of the essence of *p* that *p* is true if some entity of type E_i exists. (2) An entity, *e*, is a *truthmaker* of the proposition *p* if and only if *e* belongs to one of the entity-types E_i, which, according to axiom (1), is involved in the essence of *p*. Thus, for example, a mode, *m*, of the universal *F*ness that is possessed by the individual substance *a* would by this account qualify as a truthmaker of the proposition that *a* is *F*, for the following reason. Axiom (1) is satisfied in this case by the entity-type *mode of Fness possessed by a*, because that is an entity-type such that it is part of the essence of the proposition that *a* is *F* that that proposition is true if some entity of that type exists. And *m* is an entity of that type. Hence, by axiom (2), *m* is a truthmaker of the proposition that *a* is *F*. In effect, this is a way of cashing out more rigorously the notion of 'relevant similarity' that was toyed with earlier.

One final word: it will be noted that nowhere throughout the preceding discussion have I attempted to define *truth itself*, as opposed to the relation of *truthmaking*. This is because I take the notion of truth to be primitive and indefinable, alongside the notions of existence and identity. Only some of the family of formal ontological notions are definable and truthmaking plausibly ought to be one of them. But truth itself, I believe, is too fundamental a notion to admit of non-circular definition.

Bibliography

Aristotle, *Categories and De Interpretatione*, trans. J. L. Ackrill (Oxford: Clarendon Press, 1963).
Armstrong, D. M., 'C. B. Martin, Counterfactuals, Causality, and Conditionals', in J. Heil (ed.), *Cause, Mind, and Reality: Essays Honoring C. B. Martin* (Dordrecht: Kluwer, 1989).
——, *A Combinatorial Theory of Possibility* (Cambridge: Cambridge University Press, 1989).
——, 'How Do Particulars Stand to Universals?', in D. W. Zimmerman (ed.), *Oxford Studies in Metaphysics, Volume 1* (Oxford: Oxford University Press, 2004).
——, Replies to Fales, Menzies and Smart, in J. Bacon, K. Campbell and L. Reinhardt (eds), *Ontology, Causality and Mind: Essays in Honour of D. M. Armstrong* (Cambridge: Cambridge University Press, 1993).
——, *Truth and Truthmakers* (Cambridge: Cambridge University Press, 2004).
——, *Universals: An Opinionated Introduction* (Boulder, CO: Westview Press, 1989).
——, *What is a Law of Nature?* (Cambridge: Cambridge University Press, 1983).
——, *A World of States of Affairs* (Cambridge: Cambridge University Press, 1997).
——, Martin, C. B. and Place, U. T., *Dispositions: A Debate*, ed. T. Crane (London: Routledge, 1996).
Bacon, J., Campbell, K. and Reinhardt, L. (eds), *Ontology, Causality and Mind: Essays in Honour of D. M. Armstrong* (Cambridge: Cambridge University Press, 1993).
Bird, A., 'Necessarily, Salt Dissolves in Water', *Analysis* 61 (2001), pp. 267–74.
——, 'On Whether Some Laws are Necessary', *Analysis* 62 (2002), pp. 257–70.
Bondi, H., *Cosmology*, 2nd edn (Cambridge: Cambridge University Press, 1961).
Bradley, F. H., *Appearance and Reality* (London: Swan Sonnenschein, 1893).
Campbell, K., *Abstract Particulars* (Oxford: Blackwell, 1990).
Cartwright, N., *The Dappled World: A Study of the Boundaries of Science* (Cambridge: Cambridge University Press, 1999).
——, *Nature's Capacities and Their Measurement* (Oxford: Clarendon Press, 1989).
Chisholm, R. M., 'The Basic Ontological Categories', in K. Mulligan (ed.), *Language, Truth and Ontology* (Dordrecht: Kluwer, 1992).
——, *A Realistic Theory of Categories: An Essay in Ontology* (Cambridge: Cambridge University Press, 1996).
Davidson, D., *Essays on Actions and Events* (Oxford: Clarendon Press, 1980).
——, 'The Individuation of Events', in his *Essays on Actions and Events* (Oxford: Clarendon Press, 1980).
——, *Inquiries into Truth and Interpretation* (Oxford: Clarendon Press, 1984).
——, 'Reply to Quine on Events', in E. LePore and B. McLaughlin (eds), *Actions and Events: Perspectives on the Philosophy of Donald Davidson* (Oxford: Blackwell, 1985).
——, 'True to the Facts', in his *Inquiries into Truth and Interpretation* (Oxford: Clarendon Press, 1984).

Dodd, J., 'Is Truth Supervenient on Being?', *Proceedings of the Aristotelian Society* 102 (2002), pp. 69–86.
Drewery, A., 'Laws, Regularities and Exceptions', *Ratio* 13 (2000), pp. 1–12.
Dummett, M. A. E., *Frege: Philosophy of Language*, 2nd edn (London: Duckworth, 1981).
Ellis, B., 'Causal Powers and Laws of Nature', in H. Sankey (ed.), *Causation and Laws of Nature* (Dordrecht: Kluwer, 1999).
——, *Scientific Essentialism* (Cambridge: Cambridge University Press, 2001).
Englebretsen, G., Review of E. J. Lowe's *Kinds of Being*, *Iyyun, The Jerusalem Philosophical Quarterly* 40 (1991), pp. 100–5.
Fales, E., 'Are Causal Laws Contingent?', in J. Bacon, K. Campbell, and L. Reinhardt (eds), *Ontology, Causality and Mind: Essays in Honour of D. M. Armstrong* (Cambridge: Cambridge University Press, 1993).
Fine, K., 'Essence and Modality', in J. E. Tomberlin (ed.), *Philosophical Perspectives, 8: Logic and Language* (Atascadero, CA: Ridgeview, 1994).
——, 'Ontological Dependence', *Proceedings of the Aristotelian Society* 95 (1995), pp. 269–90.
Foster, J., 'Induction, Explanation and Natural Necessity', *Proceedings of the Aristotelian Society* 83 (1982), pp. 87–101.
Frege, G., *The Foundations of Arithmetic*, trans. J. L. Austin (Oxford: Blackwell, 1953).
——, 'On Concept and Object', in P. Geach and M. Black (eds), *Translations from the Philosophical Writings of Gottlob Frege*, 2nd edn (Oxford: Blackwell, 1960).
——, 'On Sense and Reference', in P. Geach and M. Black (eds), *Translations from the Philosophical Writings of Gottlob Frege*, 2nd edn (Oxford: Blackwell, 1960).
——, *Philosophical and Mathematical Correspondence*, ed. B. McGuinness, trans. H. Kaal (Oxford: Blackwell, 1980).
——, *Translations from the Philosophical Writings of Gottlob Frege*, 2nd edn, ed. P. Geach and M. Black (Oxford: Blackwell, 1960).
—— and Russell, B., 'Selection from the Frege–Russell Correspondence', in N. Salmon and S. Soames (eds), *Propositions and Attitudes* (Oxford: Oxford University Press, 1988).
Gracia, J. J. E., *Individuality: An Essay on the Foundations of Metaphysics* (Albany, NY: State University of New York Press, 1988).
Haack, S., *Philosophy of Logics* (Cambridge: Cambridge University Press, 1978).
Hale, B., 'Absolute Necessities', in J. E. Tomberlin (ed.), *Philosophical Perspectives, 10: Metaphysics* (Oxford: Blackwell, 1996).
Heil, J., *From an Ontological Point of View* (Oxford: Clarendon Press, 2003).
—— (ed.), *Cause, Mind, and Reality: Essays Honoring C. B. Martin* (Dordrecht: Kluwer, 1989).
Hoffman, J. and Rosenkrantz, G. S., *Substance Among Other Categories* (Cambridge: Cambridge University Press, 1994).
Jackson, F., *From Metaphysics to Ethics: A Defence of Conceptual Analysis* (Oxford: Clarendon Press, 1998).
Katz, J. J., 'Mathematics and Metaphilosophy', *Journal of Philosophy* 99 (2002), pp. 362–90.
Keynes, J. M., *A Treatise on Probability* (London: Macmillan, 1921).
Kistler, M., 'Some Problems for Lowe's Four-Category Ontology', *Analysis* 64 (2004), pp. 146–51.
Kripke, S. A., 'Identity and Necessity', in M. K. Munitz (ed.), *Identity and Individuation* (New York: New York University Press, 1971).

——, *Naming and Necessity* (Oxford: Blackwell, 1980).
LePore, E. and McLaughlin, B. (eds), *Actions and Events: Perspectives on the Philosophy of Donald Davidson* (Oxford: Blackwell, 1985).
Levinson, J., 'Properties and Related Entities', *Philosophy and Phenomenological Research* 39 (1978), pp. 1–22.
——, 'The Particularisation of Attributes', *Australasian Journal of Philosophy* 58 (1980), pp. 102–15.
——, 'Why There Are No Tropes', forthcoming.
Lewis, D. K., Critical notice of D. M. Armstrong's *A Combinatorial Theory of Possibility*, *Australasian Journal of Philosophy* 70 (1990), pp. 211–24.
——, *On the Plurality of Worlds* (Oxford: Blackwell, 1986).
——, 'Truthmaking and Difference-Making', *Nous* 35 (2001), pp. 602–15.
Locke, J., *An Essay Concerning Human Understanding*, ed. P. H. Nidditch (Oxford: Clarendon Press, 1975).
Loux, M. J. and Zimmerman, D. W. (eds), *The Oxford Handbook of Metaphysics* (Oxford: Oxford University Press, 2003).
Lowe, E. J., 'Abstraction, Properties, and Immanent Realism', in T. Rockmore (ed.), *Proceedings of the Twentieth World Congress of Philosophy, Volume II: Metaphysics* (Bowling Green, OH: Philosophy Documentation Center, 1999).
——, 'Coinciding Objects: In Defence of the "Standard Account"', *Analysis* 55 (1995), pp. 171–8.
——, 'The Four-Category Ontology: Reply to Kistler', *Analysis* 64 (2004), pp. 152–7.
——, 'Identity, Individuality, and Unity', *Philosophy* 78 (2003), pp. 321–36.
——, 'Individuation', in M. J. Loux and D. W. Zimmerman (eds), *The Oxford Handbook of Metaphysics* (Oxford: Oxford University Press, 2003).
——, *An Introduction to the Philosophy of Mind* (Cambridge: Cambridge University Press, 2000).
——, *Kinds of Being: A Study of Individuation, Identity and the Logic of Sortal Terms* (Oxford: Blackwell, 1989).
——, 'Laws, Dispositions and Sortal Logic', *American Philosophical Quarterly* 19 (1982), pp. 41–50.
——, 'Locke, Martin and Substance', *Philosophical Quarterly* 50 (2000), pp. 499–514.
——, 'Metaphysical Nihilism and the Subtraction Argument', *Analysis* 62 (2002), pp. 62–73.
——, 'The Metaphysics of Abstract Objects', *Journal of Philosophy* 92 (1995), pp. 509–24.
——, 'Miracles and Laws of Nature', *Religious Studies* 23 (1987), pp. 263–78.
——, 'Noun Phrases, Quantifiers, and Generic Names', *Philosophical Quarterly* 41 (1991), pp. 287–300.
——, 'On the Alleged Necessity of True Identity Statements', *Mind* 91 (1982), pp. 579–84.
——, *The Possibility of Metaphysics: Substance, Identity, and Time* (Oxford: Clarendon Press, 1998).
——, 'Properties, Modes and Universals', *The Modern Schoolman* 74 (2002), pp. 137–50.
——, 'Sortal Terms and Natural Laws', *American Philosophical Quarterly* 17 (1980), pp. 253–60.
——, *A Survey of Metaphysics* (Oxford: Clarendon Press, 2002).
——, 'The Truth about Counterfactuals', *Philosophical Quarterly* 45 (1995), pp. 41–59.

Lowe, E. J., 'What *is* the "Problem of Induction"?', *Philosophy* 62 (1987), pp. 325–40.
MacBride, F., 'Particulars, Modes and Universals: A Response to Lowe', *Dialectica* 58 (2004), pp. 317–33.
Martin, C. B., 'Dispositions and Conditionals', *Philosophical Quarterly* 44 (1994), pp. 1–8.
——, 'The Need for Ontology: Some Choices', *Philosophy* 68 (1993), pp. 505–22.
——, 'Power for Realists', in J. Bacon, K. Campbell and L. Reinhardt (eds), *Ontology, Causality, and Mind: Essays in Honour of D. M. Armstrong* (Cambridge: Cambridge University Press, 1993).
——, 'Substance Substantiated', *Australasian Journal of Philosophy* 58 (1980), pp. 3–10.
—— and Heil, J., 'The Ontological Turn', *Midwest Studies in Philosophy* XXIII (1999), pp. 34–60.
Melia, J., 'Reducing Possibilities to Language', *Analysis* 61 (2001), pp. 19–29.
Mellor, D. H., 'In Defense of Dispositions', *Philosophical Review* 83 (1974), pp. 157–81.
Mertz, D. W., *Moderate Realism and its Logic* (New Haven: Yale University Press, 1996).
Molnar, G., *Powers: A Study in Metaphysics* (Oxford: Oxford University Press, 2003).
Mulligan, K., 'Relations – Through Thick and Thin', *Erkenntnis* 48 (1998), pp. 325–53.
—— (ed.), *Language, Truth and Ontology* (Dordrecht: Kluwer, 1992).
——, Simons, P. and Smith, B., 'Truth-Makers', *Philosophy and Phenomenological Research* 44 (1984), pp. 287–321.
Mumford, S., *Dispositions* (Oxford: Oxford University Press, 1998).
Munitz, M. K. (ed.), *Identity and Individuation* (New York: New York University Press, 1971).
Neale, S., 'The Philosophical Significance of Gödel's Slingshot', *Mind* 104 (1995), pp. 761–825.
Olson, K. R., *An Essay on Facts* (Stanford, CA: CSLI, 1987).
The Oxford Dictionary of Physics, 4th edn (Oxford: Oxford University Press, 2000).
Plantinga, A., *The Nature of Necessity* (Oxford: Clarendon Press, 1974).
Priest, G., *In Contradiction: A Study of the Transconsistent* (The Hague: Martinus Nijhoff, 1987).
Prior, E., *Dispositions* (Aberdeen: Aberdeen University Press, 1985).
——, Pargetter, R. and Jackson, F., 'Three Theses about Dispositions', *American Philosophical Quarterly* 19 (1982), pp. 251–7.
Putnam, H., 'The Meaning of "Meaning"', in his *Mind, Language and Reality: Philosophical Papers, Volume 2* (Cambridge: Cambridge University Press, 1975).
——, *Mind, Language and Reality: Philosophical Papers, Volume 2* (Cambridge: Cambridge University Press, 1975).
Quine, W. V., *From a Logical Point of View*, 2nd edn (Cambridge, MA: Harvard University Press, 1961).
——, 'Natural Kinds', in his *Ontological Relativity and Other Essays* (New York: Columbia University Press, 1969).
——, 'Ontological Relativity', in his *Ontological Relativity and Other Essays* (New York: Columbia University Press, 1969).
——, *Ontological Relativity and Other Essays* (New York: Columbia University Press, 1969).
——, 'On What There Is', in his *From a Logical Point of View*, 2nd edn (Cambridge, MA: Harvard University Press, 1961).

——, 'Speaking of Objects', in his *Ontological Relativity and Other Essays* (New York: Columbia University Press, 1969).
——, *Theories and Things* (Cambridge, MA: Harvard University Press, 1981).
——, 'Things and their Place in Theories', in his *Theories and Things* (Cambridge, MA: Harvard University Press, 1981).
Ramsey, F. P., *The Foundations of Mathematics and Other Logical Essays* (London: Kegan Paul, 1931).
——, 'Note on the Preceding Paper', in his *The Foundations of Mathematics and Other Logical Essays* (London: Kegan Paul, 1931).
——, 'Universals', in his *The Foundations of Mathematics and Other Logical Essays* (London: Kegan Paul, 1931).
Restall, G., 'Truthmakers, Entailment and Necessity', *Australasian Journal of Philosophy* 74 (1996), pp. 331–40.
Rockmore, T. (ed.), *Proceedings of the Twentieth World Congress of Philosophy, Volume II: Metaphysics* (Bowling Green, OH: Philosophy Documentation Center, 1999).
Rodriguez-Pereyra, G., *Resemblance Nominalism: A Solution to the Problem of Universals* (Oxford: Clarendon Press, 2002).
——, 'The Bundle Theory is Compatible with Distinct but Indiscernible Particulars', *Analysis* 64 (2004), pp. 72–81.
Salmon, N. and Soames, S. (eds), *Propositions and Attitudes* (Oxford: Oxford University Press, 1988).
Sankey, H. (ed.), *Causation and Laws of Nature* (Dordrecht: Kluwer, 1999).
Shoemaker, S., 'Causal and Metaphysical Necessity', *Pacific Philosophical Quarterly* 79 (1998), pp. 59–77.
——, 'Causality and Properties', in P. van Inwagen (ed.), *Time and Cause* (Dordrecht: Reidel, 1980), reprinted in S. Shoemaker, *Identity, Cause and Mind: Philosophical Essays* (Cambridge: Cambridge University Press, 1984).
——, *Identity, Cause and Mind: Philosophical Essays* (Cambridge: Cambridge University Press, 1984).
Simons, P., 'Particulars in Particular Clothing: Three Trope Theories of Substance', *Philosophy and Phenomenological Research* 54 (1994), pp. 553–75.
Smith, B., 'Logic, Form and Matter', *Proceedings of the Aristotelian Society*, Supplementary Volume LV (1981), pp. 47–63.
——, 'On Substances, Accidents and Universals: In Defence of a Constituent Ontology', *Philosophical Papers* 26 (1997), pp. 105–27.
——, 'Truthmaker Realism', *Australasian Journal of Philosophy* 77 (1999), pp. 274–91.
Sommers, F., *The Logic of Natural Language* (Oxford: Clarendon Press, 1982).
—— and Englebretsen, G., *An Invitation to Formal Reasoning: The Logic of Terms* (Aldershot: Ashgate, 2000).
Strawson, G., *The Secret Connexion: Causation, Realism, and David Hume* (Oxford: Clarendon Press, 1989).
Tomberlin, J. E. (ed.), *Philosophical Perspectives, 8: Logic and Language* (Atascadero, CA: Ridgeview, 1994).
——, *Philosophical Perspectives, 10: Metaphysics* (Oxford: Blackwell, 1996).
Van Cleve, J., 'Three Versions of the Bundle Theory', *Philosophical Studies* 47 (1985), pp. 95–107.
van Fraassen, B. C., *Laws and Symmetry* (Oxford: Clarendon Press, 1989).

van Inwagen, P. (ed.), *Time and Cause* (Dordrecht: Reidel, 1980).
Wittgenstein, L., *Tractatus Logico-Philosophicus*, trans. C. K. Ogden (London: Routledge and Kegan Paul, 1922).
——, *Tractatus Logico-Philosophicus*, trans. D. F. Pears and B. F. McGuinness (London: Routledge and Kegan Paul, 1961).
Zimmerman, D. W. (ed.), *Oxford Studies in Metaphysics, Volume 1* (Oxford: Oxford University Press, 2004).

Index

a priori truths 4, 5, 20, 44, 107, 141–2, 153, 179, 180, 194
abstract entities 6, 76–7, 81–3, 89, 90, 98, 177, 179, 180, 183, 201
abstraction 97, 145, 168
accidental generalizations and regularities 133, 142–3, 145–6, 146, 157
acts 138, 139
actuality/actual world 137, 139, 147, 153, 156, 157, 165, 200–1
alethic monism 177, 182, 183, 185
amorphous lump, the 48, 198
anti-realism 43, 198
appearance 195
Aristotle v, 4, 5, 9, 11, 21, 25, 58, 72, 74, 78, 83, 93, 109, 123, 139, 162
Armstrong, D. M. 9 n., 10, 11 n., 12, 14 n., 16, 23 n., 24 n., 26 n., 29 n., 31 n., 45 n., 46 n., 56 n., 59, 79, 94 n., 99 n., 107 n., 108 n., 113, 128, 130–2, 134, 138, 142 n., 143–5, 147, 148–9, 157, 161–2, 166–8, 182 n., 184 n., 194, 204, 205 n.
aspects 31, 133
assertion 181
atomicity thesis 52
atomless gunk 75
attributes 16, 19, 78, 79, 82, 109, 111, 112, 113, 114, 159, 169, 170, 173; *see also* properties and relation

bare particulars 14, 27, 97, 103
Baxter, D. 168
being one of 35, 37
beliefs 177–8, 190
Bird, A. 165, 171
Bondi, H. 151 n.
Bradley, F. H. 188 n.
Bradley's regress 30, 80, 92 n., 106, 111, 166–8, 171, 204
Brentano, F. 137
bundle theories 9, 14, 26, 27, 97, 103, 123, 136, 197, 198

Campbell, K. 10, 21 n., 25 n., 81 n., 96 n., 103 n., 136, 197 n.
Cartwright, N. 173
cases (of laws) 144, 145, 160, 172
categorical properties and states 17, 60, 121–2, 134, 137, 138, 160–1

categories, ontological 5–8, 10, 34, 38, 40–4, 46, 47, 48, 57, 69, 88, 108, 109–10, 113, 196–8, 207
 fundamental or basic 7, 8, 10, 11, 12, 16, 21, 23, 39, 44, 58, 78, 103, 110, 111, 115–16
category mistake 58 n., 70
causal necessity, *see* natural necessity
causal powers 23, 159–61, 164, 172; *see also* dispositional states; powers, natural
causal relations and transactions 15, 23, 98, 100, 145, 150, 159, 193
changes 23
characterization 18, 22, 23, 29, 34, 35, 37, 39, 44–5, 46, 47, 50, 51, 58, 59, 60, 77, 80, 85, 91–3, 94, 96, 110–11, 114, 115, 125, 126, 131, 159, 162, 166, 167
characterizing ties 30, 106, 111
Chisholm, R. M. 7 n.
co-location 24
colour-properties 42, 71, 124
complexes 177–8, 183, 184, 187
composition 27, 34, 49–50, 76, 108, 185
compresence (of tropes or universals) 7, 9, 10, 14, 25, 26, 27, 97
concepts 48, 52, 56, 57, 62, 82, 83–6, 106
 formal 85
concrete entities 7, 81
conditional statements 13, 14, 17, 31, 130, 137–8, 141–2, 143, 146–7
constant conjunctions 13, 14, 28, 29, 131, 134
constants (of nature) 151–2, 157, 165, 166
constitution 34, 37, 40, 49–51
contingency, metaphysical 168, 200
contingency theorists (concerning laws) 163, 169–70, 171–2
contingent beings 206
contingent truths 205–7
contradiction 77; *see also* non-contradiction, principle or law of
copula, the 54, 71, 87, 88, 94, 95–6
cosmic coincidences or accidents 28, 134, 136, 160, 161, 162, 179
countability 49, 75, 76, 78
counterfactual conditionals, *see* conditional statements

Davidson, D. 13, 150
dependence, ontological or metaphysical 8, 27, 30, 34, 61, 72, 74, 78, 97, 108, 109–10, 112, 116, 123, 137, 169, 170, 187, 200–1
  essential 199–204, 205, 206, 207, 208, 209
  identity 34, 35, 62, 116, 117, 149, 167, 169, 199–200, 207–8
  necessary 201, 202
  non-rigid existential 34, 36, 37, 62, 116, 117, 169 n., 201
  rigid existential 34, 35, 36, 37, 50, 116, 117, 201
determinism 173
dialetheism 188–9
Dirac equation 139
dispositional predication 16, 30–2, 40, 60–1, 63, 64, 95–6, 123, 124–5
dispositional properties 121–2, 134
dispositional states 17, 18, 19, 33, 61, 79, 127, 132, 134, 207; see also causal powers; powers, natural
dispositional–categorical distinction 121–2, 124, 133, 139
dispositional–occurrent distinction 125–7, 128, 133–4, 139
dispositions, see dispositional states; powers, natural
distinctness 167, 199
Dodd, J. 182 n.
Drewery, A. 94 n., 146 n.
Dummett, M. A. E. 48 n., 55, 198

electrons 75, 135, 139, 151, 154, 166, 169–70
elements of being 43, 44, 46, 48, 59, 80, 83, 86, 111, 193–4, 195, 206, 207; see also entities
Ellis, B. 141 n., 150 n., 161, 162–3, 164
empty set, the 83
Englebretsen, G. 55 n., 64 n.
entailment 12, 29, 35, 56, 126, 143, 145, 179, 180, 184, 185
entities 5, 6, 7, 38, 39, 41, 43, 46, 47, 51, 59, 69, 111, 167; see also elements of being
essences 26, 132, 165, 170, 199, 200, 203–4, 205–6, 207, 208, 209–10
essentialism 112
  scientific 141, 148, 149–55, 163–6, 168, 173
events and processes 38, 58, 69, 80–1, 140, 150, 162
evolution 172
exceptions (to laws) 145

exemplification 19, 22, 23, 30, 31, 34, 40, 47, 79, 80, 91, 92, 95–6, 98, 111, 184–5, 187–8, 206
existence 190–1, 193
existence and identity conditions 6–7, 8, 13, 20, 26, 43, 44, 48, 49, 50, 57, 61, 75, 76, 78, 83, 88, 100, 109, 122, 149–55, 164–5
existential propositions, see propositions, existential
explanation 28
external relations 167, 171

facts 12, 13, 15, 182–8, 188, 204–5
  disjunctive 184
  negative 184
  see also states of affairs
Fales, E. 141 n., 147 n.
Fine, K. 199 n.
form and content 47–9
formal ontological predicates 193–5, 198, 199
formal ontological relations 34, 35–6, 37, 38, 40–1, 44–7, 48, 51, 58, 69, 80, 111, 114, 115, 194, 206
Foster, J. 143 n.
foundation 34
four-category ontology, the v, 15–19, 20–3, 28, 30, 31, 32–3, 39, 45, 46, 48, 58, 61, 62, 63, 65, 78–80, 93–4, 96, 100, 103, 110, 112–18, 123, 134, 159
Frege, G. 30, 48, 52, 53, 55, 56, 57–8, 62, 63, 64, 65, 82, 83–5, 106, 107, 178 n., 182 n.
functions 6, 56, 89
fundamental particles 135, 137, 139, 161, 169–70

Geach, P. T. 55, 82
generic propositions and statements 94, 144, 146
God 200, 203–4
Gracia, J. J. E. 89 n.
Great Fact, the 13

Haack, S. 181 n.
haecceities 117
Hale, B. 156 n.
Heil, J. v, 10 n., 31 n., 121 n., 133 n.
higher-order properties and relations or universals 9, 16, 17, 29, 30, 41, 42, 71–2, 79, 80, 89, 94, 107–8, 114, 117, 121, 131, 132, 133, 137, 143, 144, 162, 168
Hoffman, J. 9 n.
Hume, D. 13, 134, 136, 141, 146, 187, 188

idealism 177, 188
identity 34, 35, 37, 48–9, 50, 56, 91, 114, 138, 149, 152–3, 164, 165, 193, 199
  conditions, *see* existence and identity conditions
  of indiscernibles 8, 26
  'is' of 54, 55
  laws of 55
  necessity of 141–2, 153, 163
  sign 63, 64
  transworld 150, 153, 155, 165
imaginability 163
immanent realism (regarding universals) 25, 34, 37, 62, 98–100, 109, 114, 135–6, 144, 158, 169, 172, 200, 201, 207
indiscernibility 51; *see also* identity of indiscernibles
individuality 75, 76
individuation 27, 49, 164, 165
induction, problem of 134, 136
inherence 34, 37, 77
instantiation 21, 23, 30, 34, 35, 37, 39, 44, 45, 46, 47, 58, 59, 60, 77, 80, 85, 89, 91–3, 95, 98, 99, 110, 114, 115, 126, 162, 166
  sign 63, 64
intentionality 137
internal relations 45, 46, 59, 167
intrinsic properties 46, 59

Jackson, F. 7 n., 121 n.
Johnson, W. E. 106, 107, 111

Kant, I. 4, 5, 48
Katz, J. J. 180 n.
Kepler's laws of planetary motion 13, 29, 96, 145
Keynes, J. M. 136
kinds 5, 11, 16, 17, 18, 19, 21, 22, 26, 30, 31, 38, 39, 41, 42–3, 45, 48, 58, 59, 60, 61, 62, 77–8, 82, 83, 84, 85, 86, 92, 94, 95, 104, 109, 111, 114, 116, 123, 125, 127, 128, 129, 131, 132–3, 135–6, 144, 146, 158–9, 160; *see also* natural kinds
Kistler, M. 114 n.
knowledge 4–5
  causal theory of 180
  of laws 152–5
  see also *a priori* truths
Kripke, S. A. 141, 152, 154, 163

language 25, 55, 73, 82, 101, 124, 126, 177, 178–9, 183, 196
laws of nature 11, 13–15, 16, 17, 18, 19, 26, 28–30, 31, 32, 61, 94, 95–6, 127–8, 129, 130–2, 133, 134–6, 141–55, 156–9, 160, 163–5, 168, 169–73
  structure-based 171–2

Leibniz, G. W. 8, 26, 55, 205
Levinson, J. 14 n., 88 n., 90 n., 100 n.
Lewis, D. K. 75, 137, 179 n., 182 n., 187 n.
Liar paradox 188–9
living organisms 7, 20, 171–2
location (spatiotemporal) 24, 25, 77, 151, 196
  multiple 33, 76, 99, 158; *see also* wholly present, being
Locke, J. 82, 91, 199
logic 48, 56, 203
  quantified predicate 52–6, 62, 65
  sortal 63
  traditional formal 52–6, 64
logical necessity 141, 142; *see also* necessity, metaphysical
logical truths 202, 203, 208; *see also* necessary truths

MacBride, F. 113–18
manifestations (of powers or dispositions) 130, 137, 138, 139–40
Marcus, R. Barcan 153, 163
Martin, C. B. 10, 14 n., 17 n., 27 n., 31 n., 97 n., 121, 130, 133–4, 135, 136, 137, 139, 141 n., 182 n.
mass 149–50, 151, 152, 158
masses 7
matter 20, 75, 76, 158
Melia, J. 179 n.
Mellor, D. H. 121 n.
Menzies, P. 144 n.
Mertz, D. W. 87 n., 91 n., 92 n., 93 n.
metaphysical necessity and necessitation, *see* necessity, metaphysical
metaphysical realism 177, 188, 191
metaphysics 3–5, 8, 10, 19, 25, 69
Millikan's oil drop experiment 154
Milne, E. A. 151 n.
mind-dependence and independence 190–1
modes 14, 16, 17, 18, 19, 21, 27, 34, 38, 39, 42, 44, 45, 50–1, 78, 79, 80–1, 85, 91, 92, 95, 96, 100, 109, 111, 112, 113, 114, 116, 123, 126, 134, 159, 161, 167, 169, 170, 186, 199, 205–6, 207, 209; *see also* property- and relation-instances; tropes
Molnar, G. 130 n., 136–40
monism 177, 191, 195, 197
  alethic, *see* alethic monism
Mulligan, K. 12 n., 46 n., 130 n.
Mumford, S. 121 n.

natural kinds 20, 26, 158–9, 161
natural laws, *see* laws of nature
natural necessity 141–2, 143, 147–8, 156; *see also* nomic necessity

natural or physical science 11, 16, 180;
  see also physics
naturalism 180
natures 158, 173
Neale, S. 13 n.
necessary beings 200–1, 203
necessary connections 187–8
necessary truths 202–4, 205
necessitation relation 14, 16, 17, 29, 30, 79,
  94, 107, 131, 143–4, 148–9, 162, 168
necessity, absolute 141, 148, 156
necessity, metaphysical 26, 59, 132, 133, 142,
  148, 150, 152–3, 155, 156, 163–4, 166,
  169, 171, 173, 185, 201, 202
necessity, natural or physical, see natural
  necessity
necessity, nomic, see nomic necessity
necessity, relative 132, 148, 156
negation 53, 64, 189
Newton's law of gravitation 149–50, 156–7,
  158
Newton's laws of motion 149, 156
nihilism 198
no addition of being 46, 47, 48, 59, 80, 167
nomic necessity 132–3, 133, 141, 172–3;
  see also natural necessity
nominalism 46, 70, 102, 160, 194, 195
  resemblance 28, 197, 198
  see also particularism
non-contradiction, principle or law of 188–90,
  191, 203
non-substantial universals, see properties and
  relations
noumenalism 48
numbers 77, 81–3, 84, 89, 195–6, 197, 200–1,
  202–3

objects 6, 9, 10, 11, 12, 13, 15, 19, 20, 38, 39,
  42, 44, 45, 48, 56, 57, 69–70, 72, 73, 74,
  75–6, 79, 81, 84, 96–7, 100, 106, 114,
  116, 123, 136, 186; see also substances,
  individual
Occam's razor v, 11
occurrent predication 30–2, 40, 61, 63, 64,
  95–6, 123, 124–5
occurrent states 17, 18, 19, 33, 79, 92, 122,
  127, 134, 139–40, 207
Olson, K. R. 166 n.
omniscience 203–4
ontological categories, see categories, ontological
ontological dependence and independence, see
  dependence, ontological or metaphysical
ontological relativity 196
Ontological Square, the 19, 22, 39–40, 42,
  60–2, 78–9, 93, 114, 115, 117, 128
ontology 3–5, 6, 19, 47–9, 52, 65, 69, 70, 71,
  73, 183, 195, 196
organisms, see living organisms

Pargetter, R. 121 n.
particular–universal distinction, see
  universal–particular distinction
particularism 160–1; see also nominalism
particulars 7, 8, 9, 10, 11, 14, 17, 21,
  24, 25, 38, 39, 41, 43, 56, 57, 58, 99;
  see also bare particulars
parts 14, 27, 49, 50, 97, 108
Pauli exclusion principle 75
perception 15, 17, 23, 98, 100
persistence conditions 50
persons 85
philosophy 3
physics 19, 150, 156, 196
Plantinga, A. 142 n.
Plato 4
Platonism 25, 74, 77, 177, 186
pluralities 35, 49, 76, 82, 83, 146
point-particles 139
possible worlds 14, 108, 112, 132, 133, 139,
  141, 142, 144, 148, 149 n., 150–2, 153,
  154, 156, 163, 164, 165, 170, 172, 185,
  186, 187, 191, 202
potencies 138, 139; see also powers, natural
powers, natural 127, 129–30, 135, 136–40,
  149, 164; see also causal powers;
  dispositional states
predicables 114–15
predicates 46, 53, 54, 57, 71, 84, 87–8, 92,
  101, 102, 104, 112–13, 122–3, 124,
  183, 194
  descriptive 198
predication, 'is' of 53, 64
Priest, G. 189 n.
Prior, E. 121 n.
proper names 53, 54
properties and relations 5–6, 9, 14, 15, 16, 18,
  21, 22–3, 24, 38, 39, 42, 47, 56, 58,
  70–2, 73, 74, 82, 85, 87–8, 90–1, 97,
  100, 102–3, 116, 121–3, 126, 149–52,
  183, 192; see also attributes
property- and relation-instances 21, 25, 26, 27,
  28, 31, 32, 34, 78, 126; see also modes;
  tropes
propositional attitudes 177
propositions 12, 56, 81, 104, 105, 106, 108,
  177–80, 181, 182, 183, 185, 189, 192–3,
  202–3, 205, 207–8, 209–10
  atomic or elementary 52, 56, 105,
    106, 107
  existential 186
  necessarily true 185; see also logical truths;
    necessary truths
Putnam, H. 152 n., 154
Pythagoreanism 196, 197

quantifiers and quantification 69, 181, 195–6;
  see also logic, quantified predicate

quantum particles 75–6; *see also* fundamental particles
quantum physics 19
Quine, W. V. 47, 125, 195–8
quotation marks 181

Ramsey, F. P. (and Ramsey's problem) 43, 57, 72–4, 101–18, 168
realism, see immanent realism; metaphysical realism; transcendent realism
reality 4–5, 13, 25, 43, 108, 109, 190–1, 194, 195, 196, 197
relations 22, 49, 59, 80, 91, 102, 105, 136, 159, 160; *see also* properties and relations
relativism 177, 189–91, 195
relativity, general theory of 19
Restall, G. 186 n.
Rodriguez-Pereyra, G. 26 n., 197 n.
Rosenkrantz, G. S. 9 n.
Russell, B. 52, 53, 62, 63, 64, 65, 102, 106, 107, 108, 178 n.
Russell's paradox 71, 84, 87 n., 122
Ryle, G. 58 n.

saturatedness and unsaturatedness 30, 48, 106, 107, 206
Schopenhauer, A. 48
scientific essentialism, see essentialism, scientific
second-order properties and relations, see higher-order properties and relations or universals
semantics 25
senses, Fregean 85
sentences 178–9
set-membership 35, 41
set theory 5–6
sets 6, 35, 41–2, 81, 83, 89, 195–6
shape 124–5, 138–9
Shoemaker, S. 141 n., 151 n., 164
Simons, P. M. 10 n., 12 n., 25 n., 130 n.
Slingshot argument, the 13
Smart, J. J. C. 144 n.
Smith, B. 5 n., 11 n., 12 n., 48 n., 130 n., 186 n.
sodium chloride 152, 160–1, 163, 164, 165–6, 171, 172–3
Sommers, F. 52–6, 57, 64, 65
space 22, 25, 76, 81, 89, 98, 100
space-time 19
Spinoza, B. 19, 133
states of affairs 12, 13, 15, 56, 65, 107, 108, 109, 112, 113, 127, 128, 142, 143, 157 n., 168; *see also* facts
stereotypes 154
Stout, G. F. 102

Strawson, G. 141 n.
subjects 57, 101, 104, 112–13, 114–15
substance ontology 109–10, 162
substances, individual 9, 10, 20, 21, 22, 23, 25, 26, 27, 28, 31, 32, 38, 56, 58, 72, 78, 96–7, 103, 111, 112, 114, 123, 126; *see also* objects
substantial universals, *see* kinds
substrata 14, 18, 27, 28, 97, 103
syntax 25, 52, 54, 55–6, 58, 61, 62, 63, 65, 70, 71, 73, 82, 84, 101, 157

temporal parts 24, 69, 80
tendencies 29, 32, 131, 132, 138, 173
things 69
time 77, 81, 89, 98, 100, 191
transcendent realism (regarding universals) 25, 98
trope-theory 9, 11, 14, 18, 96, 99, 103, 136, 197, 199
tropes 9, 10, 11, 12, 14, 16, 21, 25, 26, 70, 74, 78, 81, 96–7, 109, 123, 134, 159, 161, 182 n., 186, 187, 193, 199; *see also* modes; property- and relation-instances
truth 4, 178–80, 192, 193, 202, 210
  correspondence theory of 182, 184
  ineliminability of 180–1, 182
  redundancy theory of 181
  unity or indivisibility of 4, 177, 182, 188, 191
truth-bearers 177–80, 181, 182, 192
truthmaker principle 11, 13, 71
truthmakers 31, 71, 108, 130, 142 n., 157, 167, 170, 171, 182–4, 186–8, 189, 192, 202, 204–5, 207, 208, 209–10
truthmaking 12, 157, 184–8, 190, 192–3, 202–4, 209–10

unit sets 35
unity 49, 75, 76, 82
universal–particular distinction 7, 9, 11, 17, 38, 39, 43, 56, 57, 58, 76–7, 88–90, 101–18
universals 8, 10, 12, 13, 14, 16, 18, 21, 22, 23–6, 28–33, 34, 35, 37, 41, 44, 45, 60, 62, 70, 74, 96–100, 123, 130–2, 134–6, 143–4, 149, 158–9, 167–8, 169, 186–7, 192, 193, 194, 201, 205–7
  structural 171
unsaturatedness, *see* saturatedness and unsaturatedness

van Cleve, J. 8 n.
van Fraassen, B. C. 29 n.

variables 63
victory of particularity, the 12

water 152, 161, 165–6
ways of being 14, 17, 18, 85, 90–1, 97, 182 n., 186
wholly present, being 9, 12, 24, 76, 88–9, 98–9

Wiener-Kuratowski definition of 'ordered pair' 6
Wittgenstein, L. 12, 49, 105, 106, 107, 108, 205
world, the 12, 49, 108, 205

zero 83

The manufacturer's authorised representative in the EU for product safety is
Oxford University Press España S.A. of el Parque Empresarial San Fernando de
Henares, Avenida de Castilla, 2 – 28830 Madrid (www.oup.es/en or product.
safety@oup.com). OUP España S.A. also acts as importer into Spain of products
made by the manufacturer.

www.ingramcontent.com/pod-product-compliance
Ingram Content Group UK Ltd.
Pitfield, Milton Keynes, MK11 3LW, UK
UKHW021319180426
11947UKWH00015B/1316